The Growth of Crime

Other Books by the Authors

Professor Sir Leon Radzinowicz, LL.D., F.B.A.
Fellow of Trinity College, Cambridge, England

History of English Criminal Law and Its Administration from 1750,
4 volumes 1948-1968 (Sweet & Maxwell, 1948-1968).
In Search of Criminology
(Heinemann and Harvard University Press, 1961).
Ideology and Crime (Carpentier Lectures)
(Columbia University Press and Heinemann, 1966).
Crime and Justice (edited with Marvin E. Wolfgang), 3 volumes
1971 (Basic Books, 1971).
Editor, *Cambridge Studies in Criminology*, 34 volumes (Heinemann).

Joan F. S. King, M.B.E., M.A.
Senior Assistant in Research, Institute of Criminology,
University of Cambridge.

The Probation and After-Care Service,
edited by J. F. S. King, 3rd Edition (Butterworth, 1969).

THE GROWTH
OF CRIME

The International Experience

SIR LEON RADZINOWICZ

AND

JOAN KING

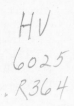
Basic Books, Inc., Publishers   New York

Library of Congress Cataloging in Publication Data

Radzinowicz, Leon, Sir.
  The growth of crime.

  Bibliography: p.
  Includes index.
    1. Crime and criminals. 2. Law enforcement.
  3. Corrections. I. King, Joan, F. S., joint author.
  II. Title.
  HV6025.R364   364   77–1958
  ISBN: 0–465–02767–9

CONTENTS

v

Part III

THE RESPONSE OF THE LAW

Part IV

ENFORCING THE LAW

Part V

THE PENAL PREDICAMENT

FOREWORD

All over the world people worry about crime and criminals. It is not only law-abiding citizens who experience these misgivings. A Mafia leader once called upon me in New York to ask why criminologists were so busy writing about the Mafia yet doing so little to prevent mugging. The ways we see crime depend upon many things, our age, sex, class, and surroundings amongst them. But everyone has views on which kinds of crime are serious and which are not. And there is a sense in which crime is genuinely everybody's business. We all suffer from it, we are all entitled to know more about it.

It is, of course, easy to exaggerate, to play upon people's fears. We are presented with successive sets of criminal statistics which seem perpetually to be creeping up: do these give us a true picture of what is going on? Or do they merely reflect more zealous policing, more accurate recording? Perhaps the rise can be accounted for by motoring offences or the myriad of minor and administrative transgressions accumulating on the statute books? Or could it be that the official figures, far from exaggerating, underestimate both the quantity and the trends of crime, leaving out of the account more and more offences that go unrecognised or unreported? If so, how many offenders are getting away? And is it true that some of them are much more likely to go scot-free than others?

There has long been a search for some fundamental means of reducing crime, whether by reforming criminals or by improving social conditions. All such schemes are now subject to critical scrutiny, if not to outright scepticism. There are those who seek the answer to crime in drastic political solutions. Have any regimes emerged in the last half-century whose prospects of containing crime seem brighter than our own?

The traditional defences against crime have been facing a barrage of criticism from more than one quarter. The charge has been made that the emphasis of the criminal law has gone wrong: is it no longer directed (if it ever was) against the evils that most threaten society; and in the attempt to regulate our complicated modern communities, has it become overstretched to a point where it can no longer be properly enforced, is no longer respected, no longer feared? Can police be efficient, trustworthy, and trusted in the face of rising crime, temptations to corruption and dangers of political involvement? Is the old symbolic role of the judiciary outplayed, to a point where extra-judicial methods of dealing with offenders are being applied on a vast scale? The place of the courts is challenged, even as guardians of individual rights. How much effort and imagination has been put into dealing with such offenders as are brought to justice, and, in particular, is it true to say that the old idea of retribution is outworn, the force of deterrence doubtful, the ideal of rehabilitation impotent?

An irreconcilable conflict is being portrayed between the call for "law and order" to keep crime in check and the call for "the rule of law" to protect individual rights. Is there indeed an unbridgeable gulf between these concepts, or are they the two indispensable sides of the coin of liberty?

Such questions are being asked in many parts of the world. There is, however, a danger in attempting to answer them in terms of one country, one situation, alone. To do so restricts and distorts our assessments and our range of responses. Today, with so much interaction between nations, so much change and questioning of where we are going, it is becoming vital to put the problems of crime and the ways in which they are tackled in some kind of international perspective.

To do this exhaustively is impossible. However, an attempt has been made in this book to take stock of the major trends and the major controversies. It is hoped that what has been garnered from records, reading, discussion, and observation over many years and in many parts of the world, may help to give a better understanding of predicaments which are by no means confined to any one country.

This book is not written for the specialist. It is intended for the much wider circle of people who would like to know more about crime and who may find, in this broader picture, a stimulus to further reflection.

Part I

WORLD CURRENTS
OF CRIME

CHAPTER
I

The Relentless Upsurge
in Crime

New Levels of Recorded Crime

No national characteristics, no political regime, no system of law, police, justice, punishment, treatment or even terror, has rendered a country exempt from crime. In fact scarcely any can claim to have checked its accelerating momentum. The incidence seems to be going up in all parts of the world, whatever the stage of development and among all segments of society, the previously law-abiding as well as the previously criminal. New forms of crime are emerging and old forms are assuming new dimensions.

England provides a good point of departure. In the year 1959 I described the criminal statistics for England and Wales as 'grim and relentless in their ascending monotony.' At that time the volume of crime recorded annually had risen from half to three quarters of a million in just over ten years. In forecasting a still sharper upward thrust, I thought that I was perhaps being a little rash. But it turned out that I had been timidly conservative. We had passed the million mark within five years and now the annual toll exceeds two million.

This rising trend has not come out of the blue. It has been gathering pace for fifty years. In all the turmoil of the present century, in the accelerating change of the past three decades, it would have been odd if crime alone had remained untouched, standing primly on the sidelines. Crime is nothing if not responsive, nothing if not opportunist. In 1900 the police of England

and Wales recorded under a hundred thousand crimes, less than three for every thousand people. In 1974 it was almost four for every hundred people. That is over thirteen times as many. And those are indictable offences, not minor infractions.

The recent speed of growth, however, has been unprecedented. It has also been peculiarly disappointing. In the first twenty years of the century, even during the first world war, rates of crime remained fairly level, no more than keeping pace with population. It was in the post-war depression that a sustained trend became discernible. Through the years of economic upheaval, unemployment and another great war, it gathered pace inexorably, though the yearly accretions were small compared with those we take for granted today. There always seemed some reason, some social evil to account for them. There always remained the hope that if we could achieve peace and plenty, they would stop. In the first half of the fifties it appeared that this hope was to be achieved. There was even a drop. We thought we were at last coming out of the wood. With prosperity, better education, full employment and the welfare state, the very roots of criminality might be starved. Then suddenly, as post-war rationing came to an end, as we were preparing to hear we had never had it so good, as the curve of affluence was gathering momentum, the curve of crime began to follow it. Its ascent during the past fifteen years accounts for two-thirds of the increase during this century. In 1974 there were over three hundred thousand more crimes than in 1973—an addition more than three times as great as the total amount recorded for 1901.

International comparison of trends in crime takes us on to dangerous ground. Different nations take different views of what is criminal, of how serious it is, and of how to deal with those who offend. What is or is not reported, the activities of police or prosecuting agencies, the definition of offences, the way they are recorded, the rules of evidence and of trial, and many other powerful factors, make their influence felt in different measure across the world. Less than a dozen countries, most of them small, can offer anything approximating to full and reliable statistics. Detailed exploration has to be confined to these. The others have only fragmentary figures, often none at all. Yet even allowing for all errors or omissions, the one thing that hits you in the eye when you look at crime on the world scale is a pervasive and persistent increase everywhere. Such exceptions as there are stand out in splendid isolation, and may soon be swamped in the rising tide.

So troubled were the Germans by their own state of crime in the sixties that they commissioned a special scrutiny of the records kept by Interpol to discover whether the malaise was peculiar to them or part of an epidemic. In France, whose police have been admired for their efficiency, if not their gentleness, since the days of the Sun King, the decade from 1955 to 1964 alone saw a rise in crime per hundred thousand population of seventy per cent. In Sweden, that model of equable prosperity, enlightened laws and penal pioneering, the rate went up forty-four per cent. In the solid and industrious Netherlands the quota increased by fifty-four percent; amongst the earnest and disciplined Germans by twenty-six per cent. No one was exempt: the Danish rate was up twenty-seven, the Austrian twenty-five and the Italian forty per cent. The pattern as well as the direction of criminal change was the same on the continent as in England: it was the property offences which accounted for most of the rise, with the strongest increases in the gravest of them like burglary and robbery. At that time it could be said that crime in England was rising faster than on the continent, but the pattern was subsequently reversed.

In the four years that followed, the crime rate in England rose little more than ten per cent, only about half the rise in Germany, Austria, Denmark, the Netherlands; only a quarter the rise in France. Oddly enough, during this period sexual offences in England were apparently increasing whilst on the continent they were declining. It is not clear whether the continentals were becoming more virtuous, or just more permissive. What is intriguing is that since then there have been signs that in this also the English are conforming to Common Market standards. What is indisputable is that new and much higher levels of crime become established as a reflex of affluence. Moreover the level is still rising. There may be years in which it is as much as twenty per cent up, others in which a rise of no more than four per cent is greeted as heralding a contraction. Occasionally one or two countries can even boast some decline. The slightest faltering in the upward climb, however transient, gives birth to too much optimism. We have learned to find crumbs of comfort in any news that, for the moment, crime is not rising quite as fast. But what matters is the general trend, and that is virtually always upwards.

If we turn westwards to the United States we meet a climate of crime far more extreme than our own. There is more of it and it is more violent. There are as many murders in Manhattan each year as in the whole of England and Wales; and Houston is even worse than Manhattan. Detroit, with much the same population as Northern Ireland, has even in these grim days five times the murders. A body of sober Commissioners, officially appointed to make a study of "criminal justice goals" throughout the United States, included in their report a veritable whirlwind of comparisons. "If New York has 31 times as many armed robberies as London, if Philadelphia has 44 times as many criminal homicides as Vienna, if Chicago has more burglaries than all of Japan, if Los Angeles has more drug addiction than the whole of Western Europe, then we must concentrate on the social and economic ills of New York, Philadelphia, Chicago, Los Angeles, and America."[1]

Moreover, the pattern of inflation in acquisitive crime has at least matched that of Europe. Homicide, rape, robbery and serious wounding all took a sharp upward turn in the last decade. In five years alone, their rate shot up by over forty per cent, twice as fast as that of offences against property. The statistics published by the F.B.I. have come under criticism and I do not deny their limitations, but great advances in precision have already been made. Because they confine themselves to the more serious crimes they offer a sober assessment of the most pressing dangers. They are, if anything, an understatement.

These upward trends are real. They cannot be accounted for merely in terms of rising population. The crime rates I have been quoting represent the numbers of offences a year for every hundred thousand people. It has been said that part of the increase in the United States can be attributed to a higher proportion of young people in the population. But the rise is much too formidable to be explained by that alone. In any case it does not apply everywhere. Nor yet can the increases be written off as simply the reflection of greater police efficiency, more crimes dragged into the daylight and written down. Could the short-lived respite enjoyed in the nineteen-fifties be attributed to a sudden police go-slow? Why has the increase affected so many countries, whether or not their police have been improved? The rise cannot be waved away, either, as simply the fruit of the massive expansion of criminal law or of motoring. The great bulk of these newer offences have been excluded from the computation: we have been looking at the explosion in the basic, old-fashioned crimes.

Only two or three of the countries prepared to back their claims with hard figures have shown signs of escaping this equation between affluence and crime. Japan, between 1956 and 1967, was able to show a large drop in homicides, a halving of frauds, a fall in robberies from five to three thousand, and even a reduction in thefts. She cheerfully attributed all this to her miraculous economic leap forward since 1955. Because of the swiftness of her industrialisation she still retains some of her pre-industrial heritage and has been trying to harness it to the new needs of urban life. Also, because prosperity came so abruptly, enthusiasm has only very recently begun to flag. There has been a powerful determination to learn from the mistakes, as well as the achievements of the older developed countries. But it remains to be seen how long all this will avail to shield her: already the omens seem threatening. Germany too can lay claim to an economic miracle, a national solidarity in the face of adversity, a tradition of paternalist concern for its people. Yet Germany has not escaped the upsurge in crime. Another exceptional case, of a different order, is that of Israel. There also I saw some evidence, both by study and observation, of a decrease, or at least no increase. Where so much of national will and of national energy, physical and spiritual, are devoted to sheer national survival, that in itself may act, for a time, as a prophylactic. But now there are signs that Israel too is succumbing to the plague.

Soviet Russia and other socialist countries find themselves in a *cul-de*-sac, constructed from their own ideology and promises. They have adopted the line that communism will destroy the roots of crime. Yet crime continues. This leaves them with only two ways out. First, they can say that they are not yet truly socialist states, or that they are still surrounded by the capitalist world, infected by its microbes, that we must wait and see what will happen when they take wing as fully-fledged socialist societies. But there comes a point when a man can no longer say that you must not expect too much of him because he is young. He may keep it up for a time, but after forty or fifty years you begin to wonder. Second, they can keep crime concealed, make statistics a state secret, as they still persist in doing. Claims that criminality is diminishing, if not vanishing, would be much more effective if they were backed by solid figures rather than occasional vague percentages.

In my opinion the socialist countries have a phenomenon of crime, a phenomenon of vandalism, a phenomenon of drunken assaults (just as in the days of the Tsars). And even if they now call theft misappropriation of Soviet property, they still have it with them, along with the white-collar crimes of the ruling bureaucracies and many other wicked things invented under the Tsarist regime—and under all kinds of regimes all over the world. On top of these they have the vast range of behaviour which can be labelled as criminal under their terrifying array of political proscriptions. Poland is a little more open about her crime. But we have to go to nonconformist Yugoslavia for comparatively public facts and discussions. She claims a decline in convictions since 1955, but much of that can be attributed to reductions in restrictive laws, to changes in her social and political situation, and possibly to the temporary emigration of thousands of her young men to work outside the country.

When I arrived recently in one of the developing countries, I found there had just been a full-scale bank robbery, as a result of which the manager was killed, his assistant wounded, and a lot of money stolen. Meeting the Minister of Justice, I said, "Excellency, I fear it is rather regrettable that my visit as a criminologist should have coincided with this bold and well planned robbery." "Not at all," was his courteous reply, "we are not embarrassed in the least. It simply means that we are becoming at last a civilised country." In some sense it is only too clear that these fast-changing societies all round the world are adding the crimes of the industrialised nations to their own traditional offences. Since none of them have anything like reliable statistics, any detailed assessment is impossible. But the very processes of development have opened the floodgates of juvenile delinquency, including the delinquency of 'protest', and of corruption and fraud.

They are particularly vulnerable, often on a vast scale, to such offences as smuggling, traffic in narcotics, illegal dealings in foreign exchange, bribery and theft in the exploitation of mineral rights or rare raw materials. It has been estimated that in Afghanistan smuggling accounts for between twenty and twenty-five per cent of all foreign trade, and that several other countries in Asia and Africa are in much the same plight. Not only do such offences involve a direct loss of revenue to the governments but they can distort whole economies, defeat any plans for constructive development. Legitimate business is involved and corrupted at many levels. Corruption is nothing new in politics, administration or commerce. But the immense scale of trade and aid in the modern world, the huge sums of money involved, have enormously extended it.

I am often asked what is going on in China: have they found an answer to crime which has eluded everyone else? Our very ignorance of that immense country makes the question as intriguing as it is unanswerable. Their passion for order and conformity is nothing new: it goes back far beyond the present regime. Yet how can we even guess, let alone measure, the nature and trends of crime in a country which has always been reluctant to define offences, let alone count them, which has always preferred the informal controls of public approval or disapproval to the formal processes of criminal law, and which quite frankly confronts us with ideological assertions of superiority, rather than hard facts? We have a few scanty clues. An acute observer has noted that the Chinese claim to have established a disciplined state, with scrupulous standards of honesty, has done more than anything else to impress intellectuals in south-east Asia. But can China, any more than capitalist Japan, hope to hold the line as she develops and opens up to the world? And if all is well why the need for continuous revolution, recurring tides of violence?

The Persisting Thread of Violence

Around the beginning of this century, though there was dispute about whether crime was lessening in quantity, there was at least a general belief that it would become milder in quality. Civilisation might not quench criminality, but it would at least transform it: the barbarous offences of feudalism would be replaced by the bourgeois offences of commercialism. An inverse proportion would reveal itself between violence and cunning. Indeed this was adopted as one of the laws expected to govern the future direction of criminality.

These hopes were the result of looking both sideways and backwards.

One ground for the optimism was found in comparisons between town and country, between the educated and uneducated, between nations at different stages of development. In the nineteenth century it could be claimed that the most industrialised nations and regions had less homicide than undeveloped areas such as Spain, Southern Italy or the frontier states of the U.S.A. And it is still possible to quote yearly homicide rates which make even those of the modern United States, now over nine per hundred thousand, look quite modest. In the nineteen-sixties Colombia reported a rate of thirty-six, Mexico of thirty-two and these rates have since gone up. In Argentina, according to recent conservative estimates, over a thousand people were being murdered every year in political strife. In less than seven years, since the outbreak of the present turmoil, the number of people killed as a result of the disturbances in Northern Ireland has mounted to 1,549, with a further 65 in England. In developing countries it is sometimes hard to draw the line between criminal homicides and the casualties of civil wars or of scrambles for political power. The classification tends to depend on who comes out on top. Yet the proposition that the stage of civilisation alone determines the proportion of violent crime gets a nasty shock when a former German concentration camp commandant is tried for murders running into thousands or even millions.

Early proponents of the 'less violence more cunning' theory relied also upon long-term comparisons within a single country. They tried to prove, on the basis of nineteenth century records, that assaults were declining even whilst thefts and frauds increased. Certain more modern criminologists, finding the statistics persistently recalcitrant, have taken a different line. It is not, they argue, that there is more violence about, but that we are much more sensitive to violence than were our less civilised ancestors. That is all very well if the comparison goes a fair way back. But what are we to think of the movement of violence in the last fifteen years? Is the whole of this to be accounted for by a sudden explosion of civilised sensibilities, a sudden desire to rush to the police for support or revenge?

A longer view, peering into the middle ages, or even the eighteenth century, might well give more substance to the theory. With all our crime, our society as a whole is more secure, less savage than theirs. But there are no statistics to guide us there. We must glean what we can from what we know or conjecture about the attitudes of people and groups from whom we are separated not merely by national frontiers but by centuries of change. The tapestries, the cartoons, the stories of violence that have come down to us are vivid enough. But this kind of criminal archaeology can be as misleading in singling out the striking and atrocious incident as it would be for a future historian to describe twentieth century society solely in terms of the moors murders or the concentration camp. It is perhaps the casual assumption, the implication about normal expectations, that gives the best clues. The mere fact that towns had to be walled, that castles had to provide refuge for the surrounding villagers and their belongings, that travellers had to take their own protection with them, bears witness to the constant threat from brigands as well as the needs of warfare. Indeed the two would often be hard to distinguish.

For all that, the fact remains that there is no law of chance or nature to guarantee that what we lose on the swings of fraud and greed we shall gain on the roundabouts of physical security. We cannot comfort ourselves, as we part with more and more of our purses, that we shall save more and more of our skins. To prophesy, or even to congratulate ourselves, in this matter is to invite contradiction. Daniel Bell, writing twenty years ago, allowed himself the luxury of pointing out that the rate of homicides in the United States had been practically halved since 1930. For good measure he added "kidnapping, the big bugaboo of the early thirties, has virtually disappeared."[2] Now people are being murdered almost as often as at the height of gangsterism, and kidnapping has entered upon a new and menacing career. Carnage on the roads far outweighs the toll of deliberate killings. In 1973 European countries listed nearly a hundred thousand traffic deaths, an increase of thirty-eight per cent in ten years. The corresponding rise in the United States was twenty-eight per cent. Of course by no means all of these fatalities can be attributed to criminal recklessness on the part of drivers, but they form a grim commentary on the idea that society, as it becomes more civilised, becomes less violent.

Now we face new threats. There is the opening up of international crime, its agents hopping across frontiers in Europe as easily as across State lines in America. Drug trafficking, always beyond the effective control of any single nation, has become a major province of criminal activity, dragging all kinds of other crime, including violence and murder, in its wake. We have had the spectacular rise of "skyjacking," which reached its peak in 1972 and 1973. The vulnerability of complex industrialised societies and the enormous destructive power of modern weapons sharpens apprehension, symbolised, at the extreme, by the haunting fear that atomic weapons might be stolen and used in criminal blackmail.

Violent crime, in the traditional sense of murder, rape, robbery, assault, produces a feeling of danger and fear on quite a different level from the most damaging offences of exploitation and fraud or the far more extensive physical injuries inflicted by negligence on the roads or at work. People crowded together in great cities are particularly vulnerable and sensitive to what has been called "random" criminal violence. One enquiry after another has been launched to assess the extent of the risk in American cities. It has been estimated that any citizen of New York has six chances in ten, during the course of his or her life, of being the victim of murder, rape, assault or robbery.

Where internecine political violence has become endemic it is not only people in cities who live under special threats. Rural farms and villages have a vulnerability of their own. As potential supply-bases and refuges for guerillas they come under double pressures: the menace and reality of violence from both the supporters and the opponents of regimes in power. In their case it is isolation rather than proximity that accentuates danger and fear.

In many parts of the world there is the violence inflicted by governments and their agencies against citizens suspected of crime or subversion. What is meant here is not the radical's claim that all law enforcement is violence but the perpetration of acts which under their own national codes would be classified as crime. Murder and torture are extreme examples. It is only rarely that practices of this kind are allowed to slip into the official records of crime, let alone be openly tried and punished. But there have been well known events in parts of Asia, Africa, South America, as well as Europe, involving the murder and torture of victims numbered in hundreds, thousands, even millions. This kind of thing is by no means past history. The impact of such transgressions is too profound to be left out of account in assessing the levels of criminal violence prevailing in the world today.

The Growing Pressures

The groups who have produced most recorded crime have been very much the same in all societies. There has been no equality of the sexes or of ages. The wicked city has lived up to its reputation. The poor have occupied a disproportionate share of places in police cells, courts and prisons. Minorities, except minorities in power, have tended to be similarly over-represented. It has been possible to explain away part of these disparities by pointing out that, at all stages of law enforcement, some groups have a better chance of escaping detection and criminal justice than others. But even allowing for this the differences have been great. What is likely to be the impact of social change in the modern world on this traditional distribution of crime? Are there grounds for believing that women's offences will increase to the point where they approach those of men, thus adding a formidable quota to the rate of criminal inflation? Are countryside and respectable suburbs getting closer to the cities in criminality? Are there reasons to hope for any check in juvenile crime? Can we detect any lessening in discrepancies between class or racial groups?

The Rise of Crime Amongst Women

Age, class, race, residence, may all be ambiguous and changing. Sex, except in the rarest instances, is not. The differences in criminality here are correspondingly clear cut, pervasive and persistent. Men and boys are found guilty six to eight times as often as women and girls. That raises two questions. First, are there any distinctive physical, physiological or psychological characteristics of women that make them constitutionally and permanently less prone to crime than men? Second, whether or not that is true, are the trends of modern life drawing women more deeply into crime than in the past?

Cesare Lombroso, never lacking in far-fetched ideas about born criminals, affirmed that women are more conservative than men, more wedded to social stability, because the ovum is less mobile than the sperm. Others, in rather similar vein, have claimed they are congenitally more submissive, conformist, adaptable, because their sexual role is passive and accepting. Comparative lack of muscular strength has been advanced, with rather less confidence, as a restraining influence: after all, there are few crimes in which strength is indispensable, and women in many times and cultures have been the principal labourers in fields and homes. More relevant, perhaps, is experimental evidence that women are, in general, less ready to take risks than men, more vulnerable to social disapproval. Even that, however, is unlikely to be a purely constitutional characteristic. Another early criminologist believed that the social position of women was at least as important in protecting them from crime as the innate modesty he thought they possessed. Their "retired and dependent state," their "sedentary life," deprived them of both the occasion and the means to compete with men in this field. Modern sociologists have attributed even more to social pressures. If women are more modest and timid it is because they are taught to submit, to be careful, to be less willing to defy the conventions. If they have committed fewer offences it has been because their life has offered fewer opportunities. They have had less chance of thefts outside their personal circle, they are seldom in a position to commit large scale frauds, they have escaped many of the occasions for assault, the brawl in the public house, the football hooliganism. Even little girls have been kept off the streets, encouraged to play at home, or left to look after the babies, much more than their brothers.

I was sternly taken to task a little while ago by proponents of women's liberation for forecasting that crime might be one of the male careers due to be invaded by women. But I had my reasons. In Germany, during the last world war, when women were obliged to take over wholesale the civilian activities and responsibilities of men away in the forces, their rates of crime came closer to those of males, reverting to the normal female level when the men came back. Amongst blacks in the United States, whose women carry heavy burdens of earning and family responsibility, the gap in crime between the sexes is narrower than amongst whites. In the poorer classes, where women are the least protected, the difference, though always there, is narrower than amongst the more prosperous. And it is narrower in the more modern societies where women share the same pursuits as men. Nowadays it is increasingly taken for granted, if only for economic reasons, that most women will go out to work for a large part of their lives, take a share in earning money to support themselves and their families. In addition we have the renewed assertion of their right to emancipation, in upbringing, in education, in lifestyle: a refusal to submit to the differential restraints that are claimed, hitherto, to have contributed to their lesser susceptibility to crime. Correspondingly, it is claimed that they are now less likely to receive special leniency at the hands of the police or the courts. In addition the modern rash of protest and revolutionary activity can be held directly responsible for an extension of female crime. Women have always been prominent in this kind of thing. During journeys in South America, I have seen girl guerillas in the front line alongside boys, their equals not only in courage but in skill, ruthlessness and resourcefulness.

Even in their domestic role, modern women in developed countries are more exposed to opportunities of crime than they used to be. Mothers have always been tempted to pilfer, sometimes in sheer desperation, to feed their families, but in the past it had to be under the sharp eyes of the shopkeeper or barrow boy. Now, in large impersonal stores, they are left to pick up what they want with every chance of concealment. At work, the old domestic servants, helping themselves cautiously to little things that might not be missed, have given place almost everywhere to women working under much the same temptations to pilfer from remote employers, or indeed to indulge in fraud, as the men alongside them. In their leisure, girls are much freer than they used to be to go their own way, choose their own company.

In the United States arrests of women over the period from 1960 to 1972 rose almost three times as fast as those of men. The trend extended not only to larceny but to such hitherto unfeminine pursuits as robbery, burglary and embezzlement. In Germany the female share in recorded crime rose, in only seven years, from 15.4 per cent to 17.1 per cent. In Canada it doubled in nine years, from 7 to 14 per cent. In Japan it rose in ten years from 9.8 to 13.6 per cent. Norway and New Zealand report similar trends.

Developing countries are not immune. Even in Moslem societies the strict seclusion which once virtually excluded women from public crime, as from all public life, is coming to an end. The new mobility of labour is taking men away from their home communities, to mines, to cities, to industrialised countries, leaving the women to carry on as best they can. India has reported that, in the four years from 1962 to 1965, the percentage increase of females amongst convicts was over four times that of males. Between 1957 and 1971 Brazil found women's convictions rising twice as fast as men's: by 89 as against 43 per cent. The development is unmistakable even if it be conceded that there is less reluctance now to arrest and convict women. A report to the United Nations Congress in 1975 on the Prevention of Crime and the Treatment of Offenders endorses the view that "most of the scant statistical evidence demonstrates that the increasing crime rates among women is a new universal phenomenon."[3]

Not all modern trends, however, can be expected to build up the share of women in crime. Some of the offences most often quoted as typically feminine are ceasing to be counted as crimes in most parts of the world. Criminologists, from Lombroso onwards, have made much play with the idea that prostitution is woman's characteristic crime, her means of getting the wealth and status for which her brothers have to steal or rob. To say that if it is not a crime it should be, or is a moral equivalent, gets us nowhere. As Baroness Wootton has neatly pointed out, each act of prostitution requires a male partner. The vast number of abortions, largely hidden, have likewise been put in the balance on the women's side of crime. But abortion, too, is increasingly being accepted as legal. A certain amount of what once constituted crime amongst men, adult homosexual offences, has likewise been relieved of the criminal label. But it is specifically women's offences that have been most attenuated by the growth of sexual freedom.

If women are not, after all, really likely to catch up with men in crime, it is because men are taking even more advantage than they are of modern trends. The reason they show percentage increases so much higher than those of their male counterparts is that they start from a far more modest base. For example, translated into absolute figures, the 89 per cent rise in the crimes of Brazilian women represents little more than a thousand extra offences, whereas the 43 per cent rise amongst men represents over fourteen thousand. To adapt Orwell, in this sphere at least, as women become more equal men become more equal still. It seems that, whatever their other achievements, women will remain, at least for a long time to come, the second sex in terms of crime. That seems particularly true at the time of life when they are most occupied in the rearing of children. It is before and after that stage that their crime is most markedly rising.

The Growing Involvement of the Young

It is one of the oldest complaints that the young, at least the male young, are prone to crime, disorder and all manner of delinquency. In England boys are up to ten times as likely to be dealt with for criminal offences as full grown men. The proportions would be broadly similar in most countries. Just as the meagre criminality of women has been attributed to their feminine characteristics, the overflowing criminality of youth has been laid at the door of childish impulsiveness or adolescent conflict. The great Belgian social statistician, Adolphe Quetelet, argued in the 1830's: "The propensity to crime must be at its maximum at the age when strength and passions have reached their height, yet when reason has not acquired sufficient control to master their combined influence."[4] In what reads like a parody of Shakespeare's seven ages of man, he ascribed petty theft to the boys, violence, rape, burglary and robbery to the young men, fraud and minor sex offences to the old, whose strength and agility were declining but who had had time to develop their cunning. The general picture, crude as it is, still holds good. In proportion to their numbers, the young claim much the highest share in violence, and the greatest increase in violent crimes. Daring, lack of foresight, uncritical enthusiasm, sheer physical strength and endurance, all play their part. You could say that these are perennial characteristics of youth in any age or clime, enough in themselves to account for a big share in mischief, small and great. In most parts of the world the young represent a much higher proportion of the total population today than even a generation ago, and this in itself contributes to higher levels of crime.

But the large role played by the young in crime, like the small role played by women, cannot be attributed to constitutional factors alone. The place they hold in society is at least as important, some would say all important. In most parts of the world that place has changed quite dramatically from what was the rule in the past. The fact that there are so many more of them does not make it any easier. In developed countries the young are kept far longer in the care, not to say the custody, of educational systems. They are absorbed far later into the now separate adult world of earning and responsibility. They tend to form a group apart, with their own aspirations, with leisure and often with spending power far beyond what was available to earlier generations of youth. On the other hand, and this applies especially to the United States, their prospects for employment are much lower than those of adults and have been getting worse. Their elders, at the same time, are probably more uncertain, more perplexed, less prepared to exercise authority, than they have been for a long time. The ancient feeling that the young of today are worse than those of yesterday persists, but there is less certainty about where the blame should be laid. Socrates could complain about young people who scoffed at authority and tyrannised over their parents and teachers. The more educated or sensitive the modern parent or teacher, the more likely he is to ask where he himself has gone wrong.

In many developing countries, under the pressures of population explosion, social, economic and political change, old patterns of responsibility for the rearing and social education of children have been dissolving even more sharply. In many parts of the world you can see hordes of youngsters, abandoned and adrift, with crime or begging their only way to keep alive. How is it possible, in situations like those, that delinquency should not increase?

The young have been even more affected than women by changes in the criminal law. Their liability to criminal prosecution has dwindled as new ideas have asserted themselves on the proper treatment of 'children in trouble.' Ever since the institution of juvenile courts at the beginning of this century there has been a growing tendency to take children, and the majority of adolescents, out of the sphere of ordinary criminal charges, criminal procedure and criminal penalties. Here it is not their activities that are classed as non-criminal but themselves. The young may be, as they almost certainly are, indulging in more damage, shoplifting, car stealing, violence, yet more often escaping conviction. Nevertheless, the criminal statistics for England and Wales showed, in 1974, that more than a quarter of those found guilty of indictable offences were under the age of seventeen, another quarter between seventeen and twenty-one. Not only that: whereas over the preceding ten years the total number of people annually found guilty of such offences had doubled, amongst those between seventeen and twenty-one it had trebled.

Gang delinquency amongst the young has attracted publicity less because of its novelty than because of its drama. After all, bands of marauding youngsters were recorded in fourteenth-century France. Shakespeare's *Romeo and Juliet* shows the youthful members of the rival families and their hangers-on roaming the streets and spoiling for a fight. In eighteenth-century Brittany complaints were recorded of "bold-faced adolescents, insolent, unbridled, mocking the laws of morality and humanity, taking pleasure in scandalising the respectable, infiltrating crowds and relieving them of their purses." Nineteenth century cities like New York, Paris or London, Bristol or Glasgow, could all produce accounts of groups of boys wandering the streets, living by pilfering from shops or stealing from passers-by. Dickens immortalised the Artful Dodger and his mates in *Oliver Twist*. People like Mary Carpenter and Dr. Barnardo tried to rescue them. Mayhew documented their background careers and activities.

International infection, stimulation, imitation, with the United States as initiator, laboratory and exemplar, has been a special feature of modern fashions in juvenile gangs, as in so much else. Gang delinquency, so called, has made its mark around the world and in all kinds of societies. The *blousons noirs* of the nineteen fifties and sixties appeared in Paris and Brussels, eventually even in stolid Switzerland. Germany had her *halbstarken*. In Italy there have been the *vitellari*, in Poland and Russia the *hooligans*, in Australia the *bodgies* and *widgies*, in Japan the *mambo*. In South Africa there has been a sub-species for each colour: the black *tsotsio*, the white *ducktails*, the coloured *skilly*. Taiwan and the Philippines had youthful groups distinguished by class: upper-crust students indulging in delinquencies connected with cars, cinemas and other modern inventions and lower-class boys going in for traditional crime. Greece, Israel, the Argentine, with their very different backgrounds, have not escaped. Even China had her Red Guards, a political instrument perhaps, but an instrument that soon threatened to get out of hand.

There is always something very menacing in the ganging-up of young men against their elders or each other, a kind of primitive fear that encourages us to exaggerate the dangers they present, the harm they do. Recently, in schools, on undergrounds, on the streets, there have been some very nasty incidents of unprovoked violence. Police have had to be stationed in the corridors of many schools in New York. Yet it is easy to exaggerate the threat. Though the tradition is persistent, most individual gangs are evanescent, mere bubbles in the current. None, in real life, have survived as long as West Side Story on stage. Take the succession in England, in only twelve years or so, of Teddy boys, Rockers, Mods, Skinheads, now the occasional bunch of Hell's Angels and the ubiquitous football hooligans—each with their own distinctive habits of dress, hairstyle, activities, outlook, emphasis. On the Continent and further afield, the young labourers known as *blousons noirs* (who set the wider fashion for a uniform of faded jeans) were succeeded by the very different *provos*, with their mixed class backgrounds, ages and sexes, their propaganda and their claims to ideological justification. It has been the same in the States: the headlines one year, obsolescence the next. And even during their brief period of life the vast majority of juvenile groups are not anything like as cohesive, as purposeful, as committed to violence or delinquency, as they are popularly pictured. Amongst the most delinquent there is far more talk than action, far more theft than violence. After spending two years in intensive observation of seven of the toughest gangs in an American slum, one investigator concluded "The average weekend of highway driving in and around Midcity produces more serious bodily injury than two years of violent crimes by Midcity gangs."

A whole range of gifted American sociologists—Albert K. Cohen, Richard A. Cloward, Lloyd E. Ohlin, Walter B. Miller, David Matza, James F. Short, Jr.—have been fascinated by the phenomenon of group delinquency. It has been the focus of

much of their effort to understand juvenile crime. Some, noting
the elements of sheer destructiveness in the boys' behaviour, have
argued that those who feel rejected by the adult world of
teachers and employers revenge themselves by reversing adult
values, smashing up property precisely because their elders set
such store by it, attacking children they think goody-goody precisely because they keep adult rules. Others have adopted the line
that such youths are simply seizing the opportunities open to
them, just as more fortunate boys take advantage of the opportunities of legitimate education and occupation. If they live in
neighbourhoods where adult crime flourishes they will imitate it
and hope, if they do well, eventually to become absorbed into
professional crime. If they live in neighbourhoods so squalid that
even the criminals are failures, they will seek an outlet in fighting
and violence. If they cannot make it either as criminals or as
fighters, they will fall back on drugs. Disappointingly, none of
these neat explanations have been verified; the search for real life
gangs that would fit them has so far failed, even in the United
States. Researchers in other countries who have tried to find
parallels report equal lack of success.

Facing the evidence that boys drift in and out of gangs, in and
out of crime, and that even the most troublesome spend most of
their time surrounded by adults and conforming to their ways,
some criminologists have produced another set of interpretations.
Far from reversing the standards of their parents, these delinquents may be reflecting the subterranean values of the adult
world or adapting as excuses the self-justifications admitted by
criminal law. Or perhaps working-class boys are merely exaggerating in their gangs the values of their own section of society,
the emphasis on being tough and smart. And possibly middle-class
boys are protesting against too much pushing around by the
women at home. We must allow sociologists to set up their im-
aginative interpretations, their sweeping simplifications, provided
they go on to test them and knock them down. It is harder to
forgive so many of them for fostering the impression, already too
widespread, that juvenile crime is produced by and concentrated
in gangs.

The fact is that the bulk of juvenile crime is committed either
alone or in twos or threes. Moreover, the group offence becomes
less and less common as boys grow older. Very few, even in the

most criminal areas, graduate from the juvenile gang to the adult criminal organisation. And it is only a tiny fraction of young people who belong at any stage to what could be classed as a delinquent gang.

Indeed in dealing with the majority of juvenile delinquents, whether solitary or gregarious, we are not struggling desperately against an intractable problem. With most it is a transient phase and will disappear in any case as they grow older. In a study which justly attracted nationwide attention, Professor Marvin E. Wolfgang followed through the careers of 10,000 boys in Philadelphia. He found that a third of them came to police attention for offences more serious than breaches of traffic regulations. But nearly half of these subsequently kept out of trouble. Only six per cent of the total proved really persistent, committing five or more offences before they were eighteen. This segment of 627 boys was responsible for over half the recorded crimes and two-thirds of the violence committed by the whole 10,000. Juvenile delinquency has been compared to measles: as a rule it takes a mild form, only for a few is the infection virulent, the complications serious. That is true in all countries: the young are especially susceptible, but the likelihood is that they will make a good recovery and bear no scars.

That, however, is only one side of the picture. It may offer consolation to parents, teachers, penologists, even to the boys themselves. But it is little comfort to the victims of burglary, arson or assault. Some of the nastier manifestations of the flare up in gang delinquency now reported in American cities should serve as a warning. The carrying and use of guns, the blackmailing of schoolmates and teachers, can hardly be accounted trivial. So reliable an authority as Dr. Walter B. Miller, investigating the situation at the behest of the Department of Justice, has estimated that New York, Los Angeles, Chicago, Philadelphia, Detroit and San Francisco have at least 2,700 gangs with well over 80,000 members, and he predicts a continued growth of juvenile gangs over the next five to ten years. By no means all juvenile crime is trifling in its consequences or costs.

Setting aside the vexed question of what should be classed as 'delinquency' amongst juveniles, and how far it should be dealt with outside the criminal law, we are left with the conclusion that people under twenty-one account for something like half of those found guilty of the traditional crimes and that if anything they have been increasing their share.

The Ethnic Component

Are blacks more prone to crime than whites, Jews or Irishmen than Germans or Englishmen, immigrants than the native born? There is a persistent tendency to regard with suspicion those of different origins, to try to assert their inferiority in intelligence, morality or both. To enter into open discussion of links between race and crime is to enter a minefield. Politically the issue is dynamite. Claims that certain ethnic groups are inferior have served, throughout history, to justify all kinds of persecution, oppression, exploitation. So powerful is the reaction now, from the oppressed and their champions, that it is hard to investigate or discuss such issues without arousing strong feelings.

This has had its effect on criminal statistics. It would be of great service to criminologists, to administrators and to public opinion in general, if national or ethnic origins, both of offenders and of victims, could be shown, along with sex and age. If differences in criminal convictions were thus brought to light, we should have a more solid foundation to allay groundless fears or as a basis for understanding and, where possible, action. But it has only been very recently, and with some doubts, that the United States has begun to provide this information. In England the idea is still rejected as a form of improper discrimination, a hindrance rather than a help in achieving the integration of immigrants. In South Africa, with a policy of apartheid, there are no such inhibitions.

On top of all that, this is a sphere in which statistics are particularly likely to be distorted by discrimination in prosecution or conviction. Many liberals in the United States are prepared to argue that the high rate of convictions amongst blacks can be accounted for almost wholly in terms of racial prejudice: police, prosecutors, courts, picking on the blacks and dealing with them more toughly than they would with whites, even whites in a similar economic position. We would partly accept this interpretation. It seems almost inevitable that groups which seem easily identifiable and are regarded with a measure of racial prejudice, are subject to greater suspicion and stronger action at each stage of the criminal process than the rest of their fellow citizens.

Paradoxically, another obstacle lies in the fact that, except in small corners of the world, the unmixed race is a myth, whether in Hitler's Germany or in the modern United States, where it has been estimated that forty per cent of the blacks are at least half white and many whites have blacks amongst their ancestors. Even supposing that distinctive racial tendencies to crime or violence were carried in the genes, it would be difficult indeed to find the pure-bred criminals to prove it. So now we talk less of race, more of ethnic and national groups.

Much confusion is brought into discussion of crime and race by neglect to distinguish the diverse social settings in which ethnic or national propensities, if any, may be displayed. You have peoples long-established in their own countries, with their own laws and history behind them. You have peoples living in their racial homelands, either as minorities like the Australian aborigines or as majorities like the South African blacks, under the domination of other races. You have peoples who have lived for centuries in a land to which they have moved or been moved, without being admitted to full equality and acceptance, as has been the case, until recently, with the Jews in Germany or the blacks in the United States. You have peoples who have very recently moved, as permanent immigrants, to a country in which they stand out as being of different race and background even if accepted in terms of rights and nationality, as is the case with Asians and West Indians entering the United Kingdom. And finally you have people going to live and work for a time in a foreign country whilst maintaining their roots in their own, like the workers from Africa or the Near East employed in Europe, or many Irishmen in England.

The early criminologists used to enjoy making comparisons between nations. Those bordering the Mediterranean, they suggested, were hot blooded and given to crimes of violence. Those whose national home lay further north were more likely to indulge in calculated crimes like theft and fraud. But it would seem that in this history and social conditions are far more important than any inborn ethnic characteristic. The Romans, at one stage of their history, had a reputation for civic virtue and the ability to rule. Where had all that gone a century or so later? The English, in medieval, Tudor and Stuart times, were a pretty violent lot, yet since the eighteenth century they have been admired for their comparative respect for order and low rates of violent crime.

Look at the vicissitudes of the Jews, a group apart, if ever there was one, in terms of their ability to retain an identity through centuries of dispersion. Have they any characteristic criminal tendencies? European studies, around the end of the nineteenth century, bore out the belief, long common in Western Europe, that the Jews in that part of the world were more likely than others to be involved in frauds and commercial offences, least likely to resort to violence, and considerably lower in general criminality than those amongst whom they lived. Their women and children were practically exempt from crime. The Nazi criminologists would have said that the Jews took to commerce in order to outwit, cheat and exploit. But, forbidden as they were for centuries to own land or enter professions, what else was open to the ambitious amongst them? Once in, they were exposed to the special temptations of an occupation in which they were particularly numerous. At the same time their very vulnerability as unpopular minorities in their countries of settlement tended to keep them close-knit. In countries and districts where they belong to the world of peasants rather than the world of commerce they play their full part in crimes of violence and theft. Modern Israel welcomes Jews from all over the world, and finds that, on arrival, they tend to reflect the standards and crimes of the very diverse lands from which they come, African and oriental as well as western, not some imagined uniform Jewishness.

Or take the blacks in the United States. Only ten per cent of the population, they account, in some years, for forty per cent of the homicides and a disproportionate share of most violent crime. But to deduce from that a special ethnic tendency to violence is quite unwarranted. If there is any weight at all in the belief that crime is produced by neglect, oppression, denial of opportunity, rejection and racial discrimination, it must be admitted that all these have borne so heavily on the blacks in the United States as to offer, in themselves, a sufficient explanation. Even the poorest whites of the slums are one step up in the social scale. Those concerned to uphold the rights of the blacks would do better to accept this than to claim that most black crime is heroic political protest. For one thing the great bulk of the violence is directed against people as deprived as they are and of their own colour. For another, the much publicised muggers are at least as likely to attack whites because they have fat wallets as because of their race. Nor is there any justification either for the reverse kind of racialism that classifies fraud, corruption or exploitation as purely white crimes, of which blacks, whatever their social position, would be incapable.

Then there are the Irish. Their rates of crime at home, except in times of political violence, have been consistently much lower, both in cities and rural areas than those for England and Wales. Yet it has been estimated that young immigrant Irishmen working in London are several times more likely to be involved in robberies and other violent offences than Londoners as a whole. Again there are many factors which could explain this, without resort to any assumption that Irishmen, as such, are especially predisposed to violence. These are young men, often from the country, suddenly away from the close control of their families and church, living largely in the most criminal areas of central London, with the temptation to seek company and consolation in pubs, where they are easily in trouble through being drunk or disorderly, fighting or picking up criminal associates. The high risks of unemployment add to the dangers. On all sides, the social situation opens the way to crime. Much the same could be said of the millions of migrant workers employed temporarily in the more prosperous parts of Europe. Cut off from their families and communities, living in societies far more sophisticated than their own, they may be especially vulnerable.

Recent coloured immigrants to England from the Commonwealth seem to have been, in general, at least as law-abiding as their white neighbours. They have been over-represented in offences related to drugs and prostitution, but that must be seen in the light of certain elements in the cultures from which they came. Violence amongst adults has been no greater than that of people of similar age, living in similar areas, especially when allowance is made for domestic rows and the pressures of housing under which many of them live. Violence amongst adolescents of certain groups undeniably exists, but it has complex roots. Cultural backgrounds, likewise, offer the best explanation for the undeniable fact that young Algerian immigrants in France appear nearly twice as violent as their French equivalents. At home they have known an emphasis on toughness, masculine dominance, pride. They are simple men, hypersensitive about French attitudes to them, and they come from a country that has endured years of violent political conflict.

What we are, as individuals and as groups, is inextricably bound up with what we have experienced and are experiencing now. All over the world national and ethnic groups with very different backgrounds and traditions are being brought into contact, living in each other's pockets. The United States has been a prime example, though certainly not the only one. Professor Thorsten Sellin, a distinguished and versatile criminologist, put forward the idea of culture conflict as part of the explanation of crime in areas containing many different immigrant groups. In its crudest form it is illustrated by the story of a Sicilian immigrant who killed his young daughter's seducer and was surprised to find himself charged with murder for simply avenging the family honour. Examples of immigrants who commit crimes because the law in their country of adoption contradicts that of their homeland have a tendency to lie on the very borderlands of crime and of credibility. It is in its subtler effects that the idea of culture conflict is most persuasive. A few groups, tightly knit and intensely loyal to their traditions, remain relatively immune and keep their children so. But it has frequently been noted that it is not the first generation immigrant who is likely to be criminal but the second. The first generation keeps itself to itself, but the children learn the ways of a new country, can become confused by two ways of life, sets of standards, may despise their parents as ignorant and become even worse than their contemporaries. If they become particularly delinquent it is at least as likely that they are reflecting the delinquency they find around them as that they are importing new forms of their own.

Prospects of reductions in crime amongst immigrant groups or those who have suffered class or racial oppression are not particularly promising. The direction of immigration is almost always from countryside to the cities, often the poorest parts of them, from less developed to industrialized countries. In other words, it is towards areas where crime is commonest. Where there has been long-standing discrimination attempts to remedy it will not necessarily pay off in an immediate reduction in lawbreaking. On the contrary, the raising of expectations which cannot quickly be fulfilled may even increase tension and crime.

Account must likewise be taken of the upsurge everywhere of sensitivity and self-assertion amongst groups which have suffered oppression and exploitation. A newly awakened racial pride in developing nations, reactions against manifestations of colonialism, abhorrence stirred up by crimes of genocide, have all reinforced this tendency in relation to race and nationality. It is true that not all the crimes committed by those suffering racial discrimination can be classed as political. But we can hardly expect any decrease in crime and violence from them at a time like this: a time of ferment, when so many social institutions are being called into question, a time when old wounds are still festering and new wounds still being received.

The Blurring of Class Differences

"Most pick-pockets, housebreakers, street robbers and footpads have once been idle vagrants," asserted an eighteenth–century Englishman.[5] "The most important function of public policy is to prevent idleness" echoed a Frenchman, "since from that fertile spring flows theft, robbery, arson, murder and almost all the disorders we see growing around us."[6]

The conviction that explanations of rising crime must be sought in the conditions of the lowest classes has persisted to this day, altering its form with changing circumstances but never disappearing. In the nineteenth century they talked quite frankly of "the dangerous classes," the urban beggars, thieves, pimps, prostitutes, vagrants inhabiting certain sectors of cities like London or Paris. Mayhew classified them wholesale under the heading of "those who will not work" and saw them as choosing to live as parasites on the rest of society, training the young to various branches of crime as to trades or professions.

As the century went on there were symptoms of a broader social conscience, a concern with the whole field of pauperism, an admission that, in Victor Hugo's phrase, such people were "savages indeed but savages of civilisation," driven out, like bands of Saxons, before their Norman conquerors, to live as outlaws on the fringes of industrialised society. Surveys like those of Booth and Rowntree at the turn of the century seemed to bear out this analysis. With so many barely able to subsist, surely crime was inevitable?

Now the old certainties are dissolving around us. Class has never been a watertight concept in the English-speaking countries: there have always been ways of crossing the boundaries even in England, incomparably more so in the United States and the countries of the Commonwealth. Today there are more. In parts of the Continent of Europe, and indeed in Asia, where class structures were once so much more rigid, they have been virtually turned upside down by revolution and social change. Now we can think of class only as an amalgam of many things: wealth, income, education, occupation, residence, in some places birth or speech or dress, in others political power or party loyalty. The modern way is to measure class by occupation and income, objective criteria in a sphere where the reality is largely intangible and subjective, bound up with the ways people think and feel about themselves, their rights, aspirations and responsibilities, and not least about their standing in the eyes of others.

Nor can it simply be assumed, as it often was in the past, that because those you see in prison come from the poorer classes these classes are the most criminal. For one thing they make up a substantial proportion of the population. Even in England, a highly developed nation, a quarter are labelled unskilled and semi-skilled. Then, too, there is often more likelihood that the man who is unemployed, or only in casual work, will be reported, prosecuted, imprisoned, rather than the skilled man in a steady job or the professional who can ill be spared. From that point of view it is not surprising if the poor find a disproportionate place in penal institutions. At the other end of the scale there is growing awareness of the extent of white-collar crime, committed by affluent professional men, business men, politicians and government officials. Corruption is essentially a persistent offence of the prosperous and the powerful, whether it takes the form of the rich bribing the rich or the powerful extorting money from the poor. Nor are the middle ranges of society exempt. The term "blue collar crime" has been coined to describe the occupational offences of workers in industry, offices and shops, of police and other officials, thus further breaking down the exclusive concern with law-breaking as a prerogative of the poorest classes, explicable mainly in terms of their conditions, physical, moral or social. We are approaching greater social equality in crime.

In a sense there have been moves towards greater equality in criminal law. The outstanding example has been the progressive removal of restrictions on the right to form trade unions, to strike and to picket. A lesser example can be found in the movement in America and England to replace prohibition by regulation in relation to the poor man's kinds of betting as well as those of the rich.

The conditions of modern life have produced a new equality of opportunity for crime. Frauds on the revenue can be committed alike by the welfare claimant and the tax evader. Not even tax evasion is the prerogative of what used to be called the middle and upper classes: it has become a way of life with many an odd-job man, builder's labourer, or small businessman. Drug abuse, like motoring abuse, seems to be a virtually classless offence. So does shoplifting in times when almost everyone goes to do his or her own shopping. Such domestic violence as baby-battering or wife-beating occurs at all levels of society. We may conjecture that the lack of full official statistics on the class distribution of various offences reflects, in part at least, the same modern embarrassment about class differences as about differences in ethnic origin.

What can we expect for the future? All over the world the passive acceptance of social inequality as pre-ordained or inevitable is giving way to demands for equal rights. But in some countries the poorest classes have been slipping further behind in income and living conditions.

A period of depression, if we have to endure it, is hardly likely to reverse the tendencies established in prosperity. What also seems inescapable is that the gradual erosion of class distinctions is unlikely to erode criminality. If anything, classes seem more likely to adopt each other's fiddles and offences than each other's inhibitions. And when we turn from recorded crime to the vast range of offences, from the petty to the prodigious, that lie concealed beneath the surface, the pervasiveness of crime throughout modern societies becomes even more apparent.

CHAPTER
2

The Expanding Dark
Figure of Crime

The recorded figures of crime are huge, but the reality behind them everywhere looms far larger. The sinister word *Dunkelziffer* was coined by a Japanese student, at the beginning of the present century, to express this hidden reality.[7] The dark figure of crime, *le chiffre noir*, *il ciffro nero*, was readily accepted into criminological currency, since it represents a very general phenomenon. Yet those responsible for law enforcement are naturally sensitive about what may be read as their failures, the yawning gap between crime committed and crime recorded, the proportion of offenders who constantly slip through the meshes of criminal justice. National pride inhibits any easy admission that it is possible, even in a well-ordered society, to get away with almost anything.

Getting Away with Murder

Murder surely stands in a class of its own, a crime of singular horror and therefore a crime that can seldom be concealed, seldom go unpunished. That is the conventional wisdom. Few countries will admit that more than one murderer in ten goes undetected. Police resources are concentrated on bringing the killer to book. It has long been assumed that this is one crime at least where the official statistics come very close to the reality. In an organised, civilised community people cannot be killed with impunity.

Or can they? The myth that murder will out is coming under fire from several directions. One or two criminologists, foremost amongst them Hans von Hentig, seem to have combed the world, like macabre stamp collectors, for the most striking instances of murderers who have carried on killing for years unsuspected. We find Max Gufler, who admitted to killing three women but at whose home were found the dentures of 'nearly eleven,' whatever that may mean. We find Grossman, who also acknowledged that he had killed three, but seven of whose previous housekeepers could not be traced despite the most diligent enquiries. Then we have a whole set of murderers who, once convicted, gloried in yet more crimes beyond the ken of the law. Dumollard was accused of killing six girls, but hinted, with cynical satisfaction, that he had been practising his terrible trade for many years; weight was lent to his claim by the discovery of over five hundred women's garments at his house, including sixty-seven stockings. Friedrich Schumann was prosecuted for six murders, eleven attempts and an assortment of rapes and thefts; in the condemned cell he calmly told his counsel, "The courts sentenced me to death six times, but I did not kill six, or even eleven, but twenty-six." John Christie admitted six murders in court, was sentenced for one, and was subsequently overheard in prison to observe that it could easily have been twelve. Even more offhand, Whitecliffe-Blume admitted twenty-seven murders but conceded, under pressure, that it was possible that he had committed four more—he really did not know, could not remember.

Nor are we allowed to imagine that all this belongs to the past. As a postscriptum comes the news of two more episodes as grisly as any of the earlier ones. In the United States and again in Mexico the discovery of a single victim has led to the exhumation of many others, people who had been killed one by one, buried and apparently never missed. Stories crop up, with grim persistence, of bodies found buried in isolated spots, of mass murders by employers or prison guards as well as by anarchists or political extremists. Mere accident can bring to light a whole train of killings that had gone on for years, concealed by the murderer and often by others as well.

You can still argue, of course, that these are the rarities. To bring them together it has been necessary to range over a century and to draw on the criminal annals of Germany, Austria, France, England and the United States. Murder is a dramatic crime and there is a temptation to seek out the dramatic in exploring its dark places, but the exceptions, the multiple murderers, do not make the rule. Nearly half of all known murders, after all, are committed by friends and close relatives of the victim. The culprit is there on the spot, often the only and self-confessed suspect. He may indeed ring the police himself, or act as his own executioner. No question of concealment can arise in such cases. Yet there is another way of looking at it. The family, as was recognised in the old law that classed murder of husband by wife as treason, can offer the ideal shelter not only for murder but for its concealment. The use of poison or drugs, the faking of suicides, are obvious examples, the very old or very young obvious targets. Recent concern about battered babies, and the slowness of hospitals to recognise or reveal that their injuries have been deliberately inflicted, indicate more violent action. Medical certificates may be based on misleading information. In a survey of deaths in English hospitals in 1958 it was found that in nearly half of them there was disagreement between the physicians who had certified the decease and those who subsequently carried out autopsies. There are too many individual cases of death originally attributed to accident or suicide but subsequently discovered to be murder to avoid the conclusion that still more exist.

Not only families but communities can shelter known murderers. They may be tribes or villages in less developed countries where killing in certain circumstances, say of twins or of witches, is counted as normal and necessary. They may be groups, whole areas, in other parts of the world where murder for political ends is classed as a praiseworthy act of war; or, alternatively, where the risk of retribution is so high as to deter anyone from going to the authorities. The silencing motives of solidarity, of resentment, of fear, can apply to murder as much as to lesser crimes.

On top of all that there is the question mark over people who just disappear. In the old days, when communications were poor and roads dangerous, there is no doubt at all that many who left their home communities simply fell by the wayside, robbed, murdered, starved. There was a great flood, soon after the siege of Paris only a hundred years ago. It floated on to the banks of the Seine more than a hundred mutilated bodies. At the same period hundreds of corpses were being found in the Thames each year. "How many victims even in peaceful times," queried a Parisian chief of police, "sleep at the bottom of both rivers whilst their murderers are still living amongst us?" Of course, in modern societies, thousands upon thousands of people disappear each year, running away from domestic or financial difficulties, perhaps just restlessly seeking a new start. But the more anonymous life becomes, the more isolated from human ties people are, the more likely it is that they can be killed not only with impunity but with few questions asked. One after another, eight elderly women, each living alone in the same street in Manhattan, were found dead during the spring and summer of 1974. Nobody saw any connection between the deaths, two of which were attributed to natural causes, two to alcoholism. When a young ex-convict was questioned about the last of them to die he told the police he had murdered them all and two others as well. He was indicted on forty-nine charges of murder, rape and robbery. It is difficult enough to get witnesses to come forward even in the most publicised cases. Some estimates go so far as to suggest that only one murder comes to light in every three or four committed, others even put it at only one in nine or ten. Without necessarily subscribing to any of these calculations, it has to be recognised that the figure must vary very widely in different societies. At bottom, it depends on the importance attached to human life, a widely fluctuating value in the world today.

Offences of Vast Obscurity

If you can hide murder you can hide almost anything. Abortion was amongst the first offences to be recognised as being both widespread and widely protected from official knowledge. It was amongst the first to invite conjectures about the 'dark figure.' Many factors tend to its concealment. It can be carried out secretly by the woman herself without the knowledge even of her family or friends. If they do know they have every motive for not reporting it. Even if she has to go to an abortionist those concerned are kept silent by complicity. Into the bargain there are all the doubts about whether abortion should be prohibited at all, all the feelings of pity and chivalry aroused by a woman in this kind of trouble. It is only when things go seriously wrong and the woman dies, or is taken to hospital gravely ill, that the offence is likely to come to light. And then any prosecution tends to be directed against the abortionist rather than his accomplice-victim.

That the great majority of illegal abortions in all countries are unknown to the police is generally admitted. The only dispute is about how enormous the hidden figure is. It has been variously estimated that there are one hundred hidden abortions for every one brought to official notice—or five hundred, or even a thousand. The figure differs from one country to another, from one period to another. In France it was calculated that before the last war there were a quarter or a third as many abortions as births, and that behind the thousand or so convictions each year lurked a hundred thousand crimes. The head of criminal police at Düsseldorf estimated that the five and a half thousand abortions shown in German police statistics for 1956 concealed something like one and a half million. More recently, less than three hundred prosecutions for abortions in West Germany in 1969 have been set against an estimate of a million illegal abortions actually taking place each year. The estimates have not been merely guesswork. Indications have been obtained from the records of hospitals and clinics, the admissions of doctors. And here again there have been the scandals that spark off enquiry. Particularly bizarre is the story of an American midwife who incurred suspicion because of the mounting rate of miscarriages in her practice, culminating in a year when more than four hundred of her patients had miscarriages, only nine normal births. All that is certain is that the police and the courts barely touch the fringe of this kind of offence.

Shoplifting is less ambiguous in its criminality, but drapes itself in almost as many extenuations and excuses. With news of shoplifting gangs and of regular, planned expeditions, we have perhaps become more cynical of late. But until recently the typical shoplifter tended to be thought of as a mother who could not make her housekeeping money go round, a woman confused by the

menopause, a pensioner craving some little luxury or a child
chancing his arm. Often it would seem heartless to prosecute
such people, even if caught. And the chances of being caught are
low. In many shops it is much more difficult to attract the attention of an assistant in order to pay than to dodge it in order to
steal. The shopkeeper himself has his motives for keeping it quiet
if he can. His own position is ambivalent, since it is his business to
tempt the customer, to allow him to see and handle the goods at
close quarters, to make him feel relaxed and at home. Research
into stealing by children in big department stores in Brussels
demonstrated that they responded, first, to the ease with which
they could pick up and handle articles, second to the feeling that
they were not being watched. In any case, the last thing many
shopkeepers want is the loss of goodwill that could arise from
publicity in court. If a thief is detected they prefer, where possible, to deal with him in their own way and recover their losses.
If the theft is not noticed in time, they write it off as part of the
inevitable cost of running a shop, either recovering it from their
insurers or adding it to the price. Two of the biggest stores in
Paris compromise with over half their shoplifters. They no
longer call in the police except for known repeaters and those
suspected of big thefts. Even then the police are obliged to drop a
third of the cases reported.

Many crimes are committed in the course of people's normal
occupations. You could say that shoplifting is an occupational
offence of housewives, though they are by no means responsible
for all of it. Then there is pilfering at work, sometimes amounting to large and persistent thefts. You get it in factories, in post
offices, on building sites, in hospitals, in hotels, on the railways,
on docks, in airports, on lorries—in fact almost everywhere that
materials, equipment or manufactured goods pass through people's hands and can be slipped out for their own use or for sale.
There is nothing new about it except the scale and the degree of
impunity taken for granted in days of plenty and high employment. In 1908 a Brussels judge could devote a book of over four
hundred pages to the peculations of *La Servante Criminelle*. Such
women, he remarked, were seldom prosecuted, merely dismissed.

The modern employer cannot so easily allow himself even the
luxury of dismissal. He may have invested a great deal of time

and money in training an employee who turns out to be dishonest. Replacing even a low paid worker may disrupt output, even provoke a strike. Like the shopkeeper faced with pilfering by customers, he may resign himself to a percentage of losses from theft, carelessness or accident, covering himself as far as he can by insurance. Faced with a flagrant example he will normally deal with it himself. Examination of the records of three big American department stores over fourteen years showed that only a tiny minority of staff caught in theft or embezzlement were reported to the police: instead they were required to make restitution. Great firms in Germany have virtually their own system of police and justice, imposing their own penalties on guilty employees without reference to outside authorities. Losses by theft may be immense but they seldom get into the criminal records.

If employers have motives for concealing offences, so have their workpeople. Feelings of solidarity, fear, complicity and personal involvement often come into it as well, alongside the feeling that helping yourself to a bit on the side is amongst the perks of the job, not a thing to go complaining about to the bosses.

Almost accidental incidents can bring to light a whole train of crimes of various kinds, committed over many years by people in high places in government and business. What seemed, at first, the commonplace capture of some incompetent burglars sparked off the immense drama of Watergate. On a far smaller scale, the bankruptcy of Poulson in England revealed ramifications of corruption. But for these incidents the offences might all have remained hidden for ever, like those bodies at the bottom of the Seine, covered up by accomplices, kept dark by the political fears of any others who stumbled across them. When the offender is armed with apparently impregnable power and respectability, the vagueness of borderlines between what is legal and illegal, complicity, anxiety about personal repercussions, can all exert added weight.

The dark figure varies according to the class of crime. It has its foothold in murder; it is considerable in the less culpable kinds of manslaughter; it is far from negligible in rape and very large in many other kinds of sexual offence, in all kinds of fraud and in robbery. In the field of theft and of more petty offences it is truly immense.

Motives for Silence

A bewildering variety of factors lies behind the process of sifting and selecting which at every stage reduces the proportion of offenders brought to justice. Some offences, by their very nature, lend themselves better than others to total concealment by their perpetrators; some offenders plan their means of eluding detection or proof as thoroughly as they plan their crimes. But very many escape through the complicity of others, not only of accomplices but the looser, multi-motivated complicity of victims, families and others who know but say nothing, at least to the police.

Commonest of all motives for silence is the feeling that it is simply not worth the trouble of going to the police. During recent surveys in the United States and Australia, more than half the victims of unreported crime (two thirds in Australia) gave as their principal reason for silence the belief that the police could not do anything about it. A further quarter put first their opinion that the police would not want to be bothered. One in four was pessimistic about the chances of catching the offender. Then there is the question of how the victim sees the offence, say an assault by a husband, a theft by a child or a friend: one in three had regarded the unreported incident as a private rather than a criminal matter. The prospect of personal inconvenience comes into it too: one in ten admitted that he had not wanted to lose time from work in attendance at court. To all this may be added those who catch an offender red-handed and prefer to settle quietly for the return of their property. And perhaps also the quasi-moral justification that we do not want the kind of totalitarian situation in which citizens make a virtue of going around reporting each other.

Involvement can be as strong a motive for silence as detachment. If we know or suspect that the culprit is one of the family, a friend or neighbour, we may not even consider going to the police. One in ten of the victims questioned in the surveys gave this as their reason, women more frequently than men, perhaps because offences against them are less likely to be committed by strangers. Rather similar considerations apply to offences within institutions or groups, whether clubs, universities or schools or delinquent gangs. There is concern for the standing of the group as well as for the individual offender. In some areas and groups hostility to the law and police may reinforce the internal solidarity. There may be the strong general tradition that "we do not talk to the police here," or the conviction that certain laws are unjust, certain penalties too harsh, and that nothing should be done to help enforce them. Compassion, chivalry, a sense of communal responsibility, may likewise go into the scales on the side of concealment. We may not want to get a child or a woman into trouble with the law, to label them, perhaps for life, as criminals. We may be sorry for the offender's family, or we may even feel it unfair that a petty thief should become a scapegoat when bigger thieves get away with it daily.

Sometimes there are the darker pressures of fear. Possibly it is fear of the aftermath if family, friends or workmates are offended by an approach to the police: the life of the sneak has never been happy. The Australian study showed women were more likely than men to keep silence for fear of reprisals, perhaps because they feel that they are at the mercy of their husbands. Or possibly there is fear of exposure: things the victim is ashamed of may be brought to light by a blackmailer, private humiliations be made public during investigations or the giving of evidence. There may even be dread of violent physical reprisals, the bullet, the fire or the bomb, a risk that obviously rises sharply in times of terrorism, war or civil war.

The motives for silence in highly organised societies, where the police are easily accessible and taken for granted, are more than matched in developing countries, with their vast rural areas. Imagine yourself farming your own patch in a remote village in Africa. Someone attacks you or steals from you. The nearest police station is many miles away, you have no telephone, no transport. The court is further away still. The time away from your crops may prove ruinous. Neither police nor court is familiar to you. You see little hope of redress even if you go to them. Their new-fangled ideas of justice take little or no account of the compensation which is, under your traditions, the chief object of bringing an offender to justice. They are too far away to protect you from his vengeance if he is punished. So why not fall back on the old, familiar ways, leave it to your neighbours or

elders to judge your case and deal with it off the record by the
old methods? It cannot be said, in such a case, that the offence is
ignored or the offender unpunished. But it means that he, like the
schoolboy, employee or club-member who is not reported to the
police, remains within the dark figure of crime as far as the
records and statistics of crime are concerned.

That is still not the end of the dark figure, however. In countries at all stages of development there are many offences reported to or observed by the police but never recorded as crimes.
A complaint of rape, for example, may be held to be unfounded
or a deliberate falsification; a complaint of theft written off as
mere lost property; a complaint of assault, especially between
husband and wife, dealt with by informal warning or conciliation. There is a vast number of minor offences going on under
the nose of the police every day, many involving no immediate
victim, hardly regarded as transgressions by most people. To
attempt to record or prosecute them all would swamp the records
and the machinery of justice. Driving too fast, getting drunk or
gambling in the wrong places, may incur no more than an informal warning. Statistics of such offences or offenders depend
almost wholly on the strictness with which the law is being enforced at a particular place and time and so are quite unreliable.
Recording may suffer still more where police are short of
men, inadequately trained, ignorant of the value of statistics,
doubtful about whether any use will be made of them.

Dramatic transformations in apparent criminality, and serious
criminality at that, have followed changes in methods of police
recording. Certain divisions of the Metropolitan Police in London
formerly provided a special 'suspected stolen book' for cases
where a theft, though suspected, had not been definitely established. Some officers used it liberally. Its abolition forced them
instead to investigate and decide in every instance whether to
record a theft or merely lost property. As a result, recorded
thefts shot up from thirteen thousand in 1931 to forty-two thousand in 1932. In 1949 the Federal Bureau of Investigation refused
to publish any more crime reports from New York, since it did
not believe the reported figures, which showed the city, twice the
size of Chicago, to have many fewer robberies. New York responded by centralising its procedure for collecting and recording complaints. This brought to light five times as many

robberies, not to mention fourteen times as many burglaries, and left Chicago far behind. In 1960 Chicago also centralised its records: its robberies thereupon overtook those of New York. In 1966, New York once more tightened its controls, and so regained the lead in honestly reported dishonesty. And the curious spiral still continues.

The understatement of crimes, in volume or gravity, may be encouraged in order to demonstrate the efficiency of a police force. That occurs in developed and developing countries alike. Over-statement may be fostered by a desire to bring home to public and politicians the need to reinforce the police, in numbers, status, equipment. Very sharp differences in recorded crime, from one year to another, or in similar areas, cannot but give rise to suspicions that what has changed is police policy in recording rather than public behaviour.

Selective Impunity: The Varying Chances of Escape

In England and Wales in 1974 nearly two million crimes were recorded but the number of persons found guilty was less than four hundred thousand.

It is thus by no means the end of the story of concealment when an offence has been duly entered in the official records. That is only the beginning of the process of bringing the offender to justice. The perpetrator may not be detected. If detected he may elude capture. If found, there may be insufficient evidence to bring him to book. Or he may be let off with a caution because of his record, his youth, his circumstances or because the offence is considered trivial. If prosecuted he may escape conviction even if guilty, or be convicted only on a lesser charge, whether through weakness of evidence, plea-bargaining, procedural protections, or a crooked defence.

The characteristics and circumstances of offenders, as well as the motives of their victims, the public or the police, play a part in deciding whether they are arrested, prosecuted, convicted. The professional robber, for example, experienced, ruthless, organised, has far more chance of escape than the amateur: even if he does not get away in the physical sense, he will make no admissions, he will know the ropes in court.

Then sex has something to do with it. Women and girls run less risk than men and boys. It is not only that their transgressions tend to be of kinds usually concealed: abortion, shoplifting, petty theft. It has often been suggested that there is more reluctance to hand them over to the police. And the English police, at least, caution sixteen per cent of females as against ten per cent of males. Before the women's liberation movement protest against this evidence of misplaced chivalry, it should perhaps be added that women are likelier than men to rank as first offenders and thus to qualify for warning rather than prosecution.

Youth is a handicap in some ways, an asset in others. The youngster is usually less capable of concealing his misdeeds. The so-called 'peak of criminality' at fourteen may stem as much from catchability as from culpability: it probably results from the maximum coincidence of the two. But being caught does not necessarily mean being reported to the police. The victim or witness may prefer to leave the family or school to deal with the delinquent, whether out of concern to save him from a 'record' or in hope that compensation can be arranged. Even before the 1969 Children and Young Persons Act in England it was nothing new for the police to caution rather than prosecute youngsters. Under the age of fourteen the odds were one in three that they would escape prosecution; between fourteen and seventeen, one in six. That compared with only three cautions for every hundred adult men.

It is debatable whether the first offender is likelier to avoid prosecution than the habitual criminal. He may be less experienced in concealment, but he is also less exposed to suspicion. There is little doubt that some who appear at court with clean records represent first prosecutions rather than first delinquencies. Certainly some who are challenged for the first time for shoplifting are subsequently discovered to have the gleanings of several establishments already stowed away in their bags or their homes. The recidivist, on the other hand, will tell you that the police are waiting to pounce whether he does anything or not, that they are on his doorstep as soon as a crime occurs in the neighbourhood. Yet there is evidence that professional criminals—even a substantial group of those classified as 'inadequate'—get away with many offences, not only before they are first convicted but in their intervals of liberty afterwards. Of over nine hundred such men subjected to intensive investigation in England a few years ago, less than thirty had not asked for additional offences to be taken into account when they had last been convicted. Two of them had, between them, committed three hundred and fifty-six offences before being caught again.

The anonymity and mobility of urban life, the disproportionate share in crime of the great conurbations, multiply the chances that the lawbreaker will be able to disappear into the crowd. However this may affect people's inclination to report to the police, there is no doubt at all about how it affects the clearing up of the crimes that are known and recorded. In England rural police forces have claimed success in forty-five cases in every hundred, urban in only thirty, with London barely topping twenty. Of course, it may be that country areas keep more of their crimes to themselves. Because there is more mutual knowledge and curiosity they can tell the police more when they choose to. But because there is more cohesion they may less often choose to. A conspiracy of silence may conceal both offence and offenders. The old smuggling and wrecking communities were classic examples in England. That kind of tight, interrelated rural enclave has few survivors in this country. But it still occurs in remoter areas of Europe, and can be the despair of legislators and administrators trying to impose modern systems of law and justice in the primitive villages of the developing world. In such situations the police may find themselves impotent even in the face of known murder.

Position in society counts. "One law for the rich and another for the poor" is a very old dictum and has been applied both to the content of the law and its enforcement. To take a few cases. A child who gets into trouble is much more likely to come before a court if he is from a very poor home and has parents who are on terms of mutual suspicion with their neighbours, the welfare authorities and the police, than if his family is prosperous and respectable, willing and able to co-operate in or arrange for his care and supervision. Again, no one doubts that theft is a crime. But the sharp practices of the rich often evade such classification. It is not only that they may be kept just this side of the law but that even if they break it they may be dealt with

outside the ordinary criminal courts. Special laws and special procedures exist in countries as diverse as the United States and
Yugoslavia. And in addition to all this the walls of wealth and
status can fend off all but the most determined investigators, the
most damning evidence, whether in averting suspicion or in com-
batting it once it arises. Some researchers go so far as to claim
that class differences lie not in crime but in impunity. A more
cautious view is that the poorer classes are probably responsible
for more than their share of crime, but that the real difference is
nevertheless exaggerated by their greater vulnerability to prosecution. Much the same has been said about blacks in the United
States.

"We're all in it together—they can't punish several thousand
of us." When it comes to the crimes of crowds the whole thing is
dramatised. The chances of impunity become visible as crime
itself assumes an extra dimension. Inhibitions are broken down by
the knowledge that windows or cars are being smashed, property
looted, unchecked by the police. Not only impunity but the
random nature of punishment becomes visible. Measures of control like gas or firing into a crowd are no respecters of persons.
The old military practice of decimation—executing every tenth
man in a mutinous regiment—acknowledged and regularised this
impossibility of punishing everyone or even of identifying the
most guilty, when a large body of people is involved. People who
think they can get away with anything in the thick of a mob may
be right. On the other hand it may go the other way: they may
incur not less but more punishment than they ever deserved.

Probing the Darkness

During the past two or three decades criminologists, especially in
the United States, Scandinavia and Germany, have been making
systematic attempts to measure and analyse the dark figures of
crime, the proportions of people who have, locked away in their
past, offences quite unknown to the authorities or, if known,
passed over.

There are several ways of trying to find out. Groups of ordinary people (in practice it has usually been schoolchildren, students or military recruits, since these are the easiest to get hold of) have been asked to indicate, in the strictest confidence, whether they have been guilty of certain actions which, if discovered, could have brought them before the courts. The offences specified have ranged from arson, theft, personal violence, vandalism, to truancy or smoking or drinking whilst under age. The youngsters have been asked to complete, quite anonymously, a written questionnaire. Some investigators claim that this method is likeliest to elicit the truth since it is the most impersonal, least embarrassing: others prefer a personal interview. Sometimes the subject is given a pile of cards, each describing an offence, and asked to sort them so as to indicate which he has committed, and how often. Questionnaire or sorting may be followed by interview so that mistakes can be cleared up, omissions remedied, trivialities excluded. Honesty may be checked by comparison of confessions with information gleaned from police, criminal records, or associates. One written questionnaire was followed both by an interview and a lie-detector test, with the result that all those questioned had second or third thoughts and changed their original answers.

Obviously there must be reservations about this kind of approach. Such researches cannot cover the whole gamut of offences, and they often include quite trivial delinquencies, such as truancy. The investigator who thinks he can elicit the same searching of the past and the conscience as the priest or the psychiatrist deludes himself. It is he who is the client, who is seeking help, and there is no reason why those he approaches should try very hard to make their answers accurate or complete. They may exaggerate, they may conceal, they may be overconscientious, exhibitionist, above all forgetful. Besides, most people soon lose track of their trespasses, or play them down or embroider them, and they have very varying ideas as to which should be counted as breaches of the criminal law. The questioner is about as likely to get an exact picture of the frequency and gravity of their offences as Dr. Kinsey was to persuade women to confide to him all the details of their sexual experience.

Even if something is discounted, however, the findings remain startling, particularly as they have been very similar in different countries and situations. On the basis of their own admissions, it

appears that very few people—less than one in ten—have never
been guilty of lawbreaking at all. And several of the studies show
that well over half confessed to at least one crime for which, if
convicted as adults, they could have been sent to prison. Of these
people only about one in ten may have been brought to book,
and not more than three in every hundred of their crimes are
known to the police. On the other hand there is some evidence
that whether or not a delinquent comes to the notice of the law is
not purely a matter of chance. It emerged from Dr. Donald
West's long-term follow-up of a group of juveniles, that boys
brought before the courts, and especially the recidivists amongst
them, tended from the outset to have differed significantly from
their peers in both social and personal respects. Official convictions tended realistically to pick out the worst behaved boys.

The method of direct observation has likewise demonstrated
that apparently law-abiding people of both sexes and from all
walks of life will break the rules, given enough temptation and a
good chance of getting away with it. For example, one in every
fifteen customers shadowed round big stores in the States—one in
twelve in New York—was seen to take something without paying. And out of a hundred of these thieves only one was apprehended by the large staffs of store detectives.

An obvious alternative to questioning or watching the perpetrators is to ask the victims. Again the most accessible have been
the first to be probed: institutions and businesses which might
keep their own records of thefts by their employees even though
they did not report them to the police. Germany has seen a series
of fascinating investigations of crime-concealment by big organisations. In the Post Office it was found that less than half the
thefts committed were tracked to their perpetrators and even then
a third of the offenders were not prosecuted. In the electrical
industry a third of the employees guilty of offences were merely
cautioned, a third dismissed, only a tenth reported to the police.
Then there are the kinds of business that are robbed not only by
their employees but by their customers.

Like questioning people about their own delinquencies, this
rather haphazard investigation of losses and concealment in particular organisations serves to confirm that hidden offences are
extremely widespread. But it gets us little nearer to actually
measuring the width of the gaps between crimes committed and

crimes officially recorded, between the reality and the statistics. It is only in the last ten years that researchers have begun to get down to this, first in the United States, then in Australia and in England. People have been interviewed to discover whether they, or their families, have been the victims of crime within the preceding twelve months and whether they told the police about it. It is possible to compare the hidden offences thus uncovered with those listed in the official criminal statistics for the same periods, as a first step in coming to grips, for the first time, with the problem of actually measuring the dark figures.

The pattern of findings, as might be expected, is variable. New York and Sydney, for example, turned out, on this basis, to have at least twice as much crime as was reported; Los Angeles, Chicago and Detroit three times as much; Philadelphia, whose records have, in the past, aroused suspicions of deliberate falsification, five times as much. But the general picture is similar. Something like twice as many burglaries and major thefts; two, three or four times as many robberies. In the United States they uncovered five times as many minor thefts as were recorded. In Sydney they found, in addition, almost eight times as many cases of mischief and arson, nine times as many sex offences short of rape, and thirteen times as many assaults.

The launching of these enquiries, supported by departments of justice, is to be welcomed. So is the decision to put them on a regular footing and publish them in parallel with the statistics of offences recorded by the police. They have special value in reflecting not only the prevalence of crime but the public response to it. Their results have been startling. But I am convinced that they are still underestimates. Many of the reasons which keep people from reporting crimes to the police will also influence them here. In addition there is evidence that, in response to questioning later, they are less likely to exaggerate than to forget or omit. There remains a long way to go in probing the extent of hidden crime.

I would give considerable weight to the estimates of those who have spent many years in the administration of criminal justice: police, prosecutors, lawyers, judges. Their assessments make no claim to exactness, but they have the feel of the thing, an intuition based on accumulated experience. The experience is inevitably biased, the intuitions may be jaundiced, but they come straight from the horse's mouth and I find them more persuasive than most. Such an estimate, based on statistics of recidivism and analyses of individual cases, was made in the nineteen-fifties by a German Chief of Police, published under the authority of their Central Office of Crime Detection and never contradicted since. Indeed the latest publication issued on behalf of that office, dealing especially with economic crime, seems to more than confirm his modest estimates. He set the ratio of the unknown to the known in Germany at five hundred to one for abortions and homosexual offences, between five and ten to one for rapes and indecent assaults upon children, between three and six to one for homicides, eight to one for frauds, and one to one for thefts, burglaries, robberies, extortions.

Naturally, these estimates, however near the mark for Germany at the time, cannot be applied automatically to other societies, since the most important factors which go to make up the dark figure vary enormously from country to country. But if, in order to illustrate the impact of hidden crime, we were to extend them to England and the United States, we should find the total of offences more than doubled: over four million indictable offences for England instead of two million. By the same criteria the United States would have over ten million serious crimes every year, as compared with the five million recorded, including over forty thousand homicides, nearly two hundred thousand rapes, and four million burglaries.

Alongside the estimate that well over half the crimes committed never reach the criminal statistics can be set the certainty that well over half of the crimes recorded are not cleared up. That leaves considerably less than a quarter of offences and offenders in the open. But even that fraction is reduced when you take account also of decisions not to prosecute, of charges reduced below the level which the offence really warrants, and of failures to convict the guilty.

In half a lifetime of watching and comparing, I have found myself inexorably driven to the conclusion that we have been much too modest in our estimates of what is hidden. In the nineteen-fifties I was given a sharp thrust in that direction by evidence brought to light by two of my colleagues at Cambridge in the course of a systematic study of sexual offences. That led me to express the doubt whether more than five in a hundred of such offences (leaving aside the serious crime of rape) were ever likely to come to light. Eventually I felt forced to the conclusion that the picture was almost as obscure for crime as a whole. In a lecture to the Royal Society of Arts in 1964, I went so far as to suggest that crimes fully brought into the open and punished represented no more than fifteen per cent of the great mass actually committed. I admitted at the time that it was only a guess. Some of my colleagues have been so kind as to call it an informed guess. Others, less graciously, have dismissed it as pure speculation. Yet others have objected that guessing is easy: it is hard, accurate evidence that is difficult to come by. On that I would agree with them. But the evidence has lately been mounting, and I stick to my estimate.

The Proliferation of Unrecorded Crime

Nor can I see any justification for claiming, or hoping, that the gap between crimes hidden and crimes recognised, criminals who escape and criminals brought to justice, has been narrowing. Or indeed that it is likely to narrow. On the contrary, there is much to suggest that, taking crime as a whole, it is becoming wider. In the last century it was believed that by making laws more moderate, and so more enforceable, by growing social awareness of crime, by establishing professional police forces, the chances of impunity would, if not eliminated, be greatly curtailed. That has not proved to be the case. Indeed I am inclined to believe that the criminal of today is more likely to remain hidden than his predecessor of some forty or seventy years ago. The evolution of society, demographic and social, seems to favour him in this, as in so many other ways. The dark figure tends to increase as society becomes more urbanised and anonymous, and also as it becomes more mobile. Criminals who, after all, are part of our society, share these characteristics.

That cuts across the consoling theory that the swelling figures of recorded crime are being replenished by dipping deeper into the reservoir of concealed offences rather than from any real increase in criminality. It may hold good for a few kinds of crime, in particular for minor assaults. We are more sensitive about our persons than we used to be. But for crime as a whole both the trends in public attitudes and the strains being felt at all stages of criminal justice point towards more concealment.

Several streams of public feeling converge to reduce still further our willingness to report offences, even whilst we clamour for stronger action by the police and courts. Especially since the last world war we have been less concerned with abstract ideals or public duty, being perhaps both lazier and more permissive. The groups christened activist are precisely those prepared to fight authority rather than co-operate with it. Greater understanding of the social influences upon individuals and upon crime has contributed to feelings of compassion, anxiety to understand rather than condemn out of hand. At the same time the whole administration of the law and its penalties have become milder, there has been less chance of satisfying cravings for revenge, retribution or strict justice. On a more mundane level, the delays and frustrations of the law have multiplied as crime increased. The chances of recovering stolen property get lower. Theft itself, the most prevalent offence of all, loses some of its impact in times of plenty. The situation in the prosperous west has become very different from that in peasant communities living on the edge of starvation, where a thief may be beaten almost to death before he even gets to the police.

Not all offences are likely to have been equally affected by these trends, or affected in the same ways. The spread of insurance has increased the likelihood that substantial thefts will be reported, if only to back a claim for the loss. But that does not apply to the same degree to the mass of minor thefts. It is these crimes that are most likely to be considered excusable if the offender is known, not worth bothering about if he is not. It is these crimes that are likeliest to be considered excusable. Minor sexual offences, like indecent exposure, may likewise be less often reported in these less shockable days. Where the law itself has recently been relaxed, as in England with homosexual offences and abortions, the effects on the dark figure are very hard to judge. A great many actions that would formerly have been classed as hidden crimes have simply ceased to be crimes at all. Those that remain criminal, however, may be considered the most serious of their class: the homosexual seduction of the young, or the use of force, the abortion carried out by the un-qualified back-street practitioner. Has the change in the law meant that a higher proportion of these cases are reported and prosecuted? Or has it simply meant that most people have come to regard both classes of behaviour as outside the proper concerns of the criminal law and the police? Only in offences of personal violence, no more than five per cent of the whole in developed countries, is there evidence that some of the rise in the official figures can be explained by greater public inclination to complain to the police. And even here, with the growth of civil strife and intimidation, counteracting tendencies are becoming more powerful.

On top of the pressures upon the public to keep quiet about offences, there are other pressures on the police and the criminal justice system. Not only is a great deal more crime concealed, a great deal more has, nevertheless, been reaching official attention. And there seems to be a kind of equilibrium between the volume of crime reported and the proportion of it that can be dealt with effectively by a given strength of police. In the matter of detec-tion, in spite of a few recent tactical victories at selected points, the police have been falling behind everywhere and in every kind of offence. That applies not only to minor thefts, which they may feel they have little time to attend to, but to crimes like murder, homicide and robbery. To put it bluntly, the more crime there is, the better the chance of escaping detection.

As for decisions to prosecute, whether taken mainly by the police, as in England, or by public prosecutor or examining magistrate, as in the United States or on the Continent, the pres-sures are also apparent, in a growing resort to cautioning, or the dropping or reduction of charges. Statistics of formal cautioning have been available in England only since 1955, but they show that by 1965 they had already expanded from eleven to fourteen per cent of people admitting indictable offences. In France a century ago only a third of offences were eliminated from the judicial process as trivial, unsubstantiated or undetected. Now the proportion is two-thirds. In the United States, of offenders taken into custody one third are not prosecuted at all and another third are charged with lesser offences as a result of plea-bargaining between prosecution and defence. It is not suggested that these various processes of elimination or reduction are necessarily or always unjustified or improper, but their general increase, in proportion to the cases coming forward, points strongly to a striving towards balance between the level of crime and the formal processes of justice which, as crime becomes more prevalent, must inevitably allow a larger proportion of offenders to escape.

In war, in crisis, in places where society is deeply divided, or where a new society is being built up from scratch, the bulk of unrecorded crime and the capriciousness of recording are far greater. Such situations may be exceptional in England, Switzerland or Sweden, but in many parts of the world they are commonplace. There the dark figure of crime, the chances of impunity, can reach fantastic heights, bobbing up and down with the vicissitudes of licence and control, by the police, by armies, by anyone who can seize power.

Some Unanswered Questions

So fundamental a phenomenon as the dark figure of crime cannot fail to affect the whole area of delinquency, as well as our awareness and understanding of it. Here are some of the more significant implications.

First, the dark figure makes it extremely difficult to ascertain the true composition and trends of criminality. Quetelet, the Belgian who fathered the statistical analysis of crime in relation to society a century and a half ago, was too acute to overlook this problem, but too determined to let it stand in his way. He got round it by the simple assumption that, so long as social conditions remained the same, the proportions between recorded and hidden crime would remain constant. On that basis the official statistics could be used as a convenient indication of what was happening for crime as a whole. Of course there is a fatal flaw in this. The one thing modern society does not do is remain constant. And the dark figure can alter in different directions and at different rates for different offences. All attempts to interpret relations between crime and social change need to take account of that.

Second, the dark figure restricts and distorts our knowledge of

Part II

EXPLAINING CRIME

CHAPTER

3

Ideologies and Crime

The attempt to understand crime has a long history. We cannot fully comprehend the various attitudes prevalent today without reference to the cultural and social conditions which gave them birth in the past. Looking back over time, we can identify a series of constellations of ideas about crime and criminal justice. Often they have been taken for granted, implied rather than explicit. Yet, particularly in some countries and at some periods, they have been brought to the surface, consciously crystallized and expressed.

Where an explicit and coherent set of attitudes is shared by academic lawyers, judges, administrators and leaders of public opinion, we can speak of a "school" of criminal law or criminology. The Continent of Europe, where academics have been particularly influential in formulating attitudes towards crime and criminals, has seen a succession of rival schools of this kind. Until recently England and the United States, where legislatures and judges have played a far more decisive role, have stood aside from these controversies. That has, indeed, been true of the whole English-speaking world. These countries have taken a more pragmatic line, responding to the pressures of events, meeting the problems of crime as they came up. Now they too are being affected by the opposition between fundamentally different approaches to the understanding and control of crime, the nagging demands to have done with compromise and commit themselves to a single unambiguous attitude.

Liberal Approach

The first modern school of criminal law emerged against a background of arbitrary severity in eighteenth-century Europe. The 'enlightenment' in Europe brought a revolt against the unquestioning acceptance of tradition and authority, protests against the pervading cruelty, appeals to what were seen as 'natural law' and 'natural rights,' demands that the law should protect the citizen and his liberties. The liberal doctrine of criminal law, as summed up by the young Italian Cesare Beccaria, challenged, one by one, the assumptions and practices of arbitrary systems. These were its central tenets. First, the law should restrict the individual as little as possible, it should guarantee the rights of the accused at all stages of criminal justice, it was no part of its function to enforce moral virtues as such. Laws not essential to serve the needs of a particular society uselessly increased the tally of crime. Second, it should guarantee the rights of the accused at all stages of criminal justice. Third, it should give clear and precise knowledge of what was forbidden and the sanction attached to disobedience. Fourth, it should take the form of a complete written 'social contract,' so that people could judge how their liberties were being protected. Fifth, punishment was justified only in so far as the offender had infringed the rights of others. Sixth, its severity should be drastically curtailed: it should be no more than proportionate to the crime committed, and it should not go beyond what was necessary to deter the criminal and others from injuring their fellows. It should be just sufficient to ensure that the penalty outweighed the advantage derived from the crime. Seventh, the nature of the penalty should correspond with that of the crime: fines would suffice for petty theft, but corporal punishment or physical labour should follow crimes of violence. Imprisonment seemed a particularly appropriate penalty in an age that valued liberty above all; moreover, it lent itself well to the most exact gradation between crime and punishment. Eighth, punishments must be inflicted with speed and certainty, in order to create the closest possible association in people's minds between crime and its inevitable penalty. This would be easier to achieve if sanctions were seen as just and moderate. Ninth, exemplary punishment must be ruled out as unjust. Likewise 'reformation' must not be thrust upon criminals: they should not be compelled to undergo anything except the legal punishment. And, since the punishment was to be strictly related to the crime, it should not be varied to suit the personality or circumstances of the offender. He should suffer for what he had done, not for what he was or might become, or for danger he might threaten in future. Tenth, the criminal was to be treated as a rational responsible being, who could weigh up the consequences of crime and had the same power as anyone else to resist it. Eleventh, the prevention of crime was to be achieved by the clear code of offences and punishments, supplemented by better education and better rewards for virtue: in the light of their experience the liberals of the enlightenment utterly rejected the idea that crime could be or should be forestalled by police watchfulness. The mainstay of prevention must be not the regulation of morals, nor control of suspects, but punishment of actual crime.

The principles of this doctrine were never fully translated into reality. But thinking, writing and teaching were profoundly inspired by them. Even more important, new codes of criminal law and procedure all round the world aspired to follow the liberal creed. The process spanned most of the nineteenth century, from the Declaration of the Rights of Man and the first criminal code of the French Revolution, by way of the Code Napoléon of 1810, the Bavarian code drawn up by Feuerbach in 1813 and the Zanardelli code for Italy of 1889. Even in England, with its suspicion of cut and dried systems, the abortive yet influential code constructed by Sir James Fitzjames Stephen, like Edward Livingston's famous "System of Penal Law" for the State of Louisiana in 1824 and the code drawn up for New York by David Dudley Field in 1865, all bore the unmistakable marks of this same influence. What had begun with the liberal and revolutionary impulses of the enlightenment acquired, in the sphere of criminal justice, the title of the classical school of criminal law and penal policy. Classical it certainly was in its apparent simplicity, its idealism, its emphasis on proportion and restraint, on logical rules to be known and accepted by all.

Positivist Approach

The first major onslaught upon the elegant structure of classicism
reverberated in its turn all round the world. Launched towards
the end of the nineteenth century, it was again the product of a
European school, based on radical criticism of the immediate past
and projecting a revolutionary programme for the future control
of crime. Though schools are inevitably identified with their
leaders, their conception, growth and acceptance depend upon
the environment of their time. The liberal outlook had been born
with the assertion of human rights, and the classical school had
developed in response to the need for a reformed and stable
order. But the new wave of reformers had but a dim perception
of the arbitrary abuses that had impelled liberals and classicists to
these ends. They were living in a Europe where it seemed stability could at last be taken for granted, and with it freedom from
the tyrannies of the past. What concerned them was progress
towards greater social justice.

The classical school, one of whose leaders instructed his pupils
that crime was not an entity in fact but an entity in law, showed
no interest in studying the sources of criminal behaviour. But
with Darwin opening up the natural sciences, with Comte and
Spencer establishing the social sciences, with the beginnings of
modern psychiatry and psychology, what was more natural than
that all these should be extended into the explanation and control
of criminal behaviour, whether seen as part and parcel of society
or of the individual human personality?

When the classical school emerged the measurement of crime
was in its infancy. But by the time the new school was born
several European countries could offer criminal statistics which
made it possible to compare the trends of crime over substantial
periods, to recognise its persistence, to challenge the failure of
classical doctrines to bring it into check and to claim that a
totally new approach was essential for the protection of society.

Three major books, all the work of Italian scholars, heralded
this second revolution in attitudes towards crime. The first was
L'Uomo Delinquente, The Criminal Man, by Cesare Lombroso
(1876). The second was *Criminal Sociology* (1878) by Enrico

Ferri. The third was *Criminology* (1884) by Raffaele Garofalo. These three and their disciples adopted with pride the title of positivists, the term used by Comte to define his own system of philosophy. Like him, they insisted that what mattered was not idealistic speculation but positive facts and observable phenomena. They emphasised experiment rather than deductive reasoning. They might well have taken as their maxim the words of the veteran French statistician, André-Michel Guerry: "The time has gone by when we could claim to regulate society by laws established solely on metaphysical theories and a sort of ideal type which was thought to conform to absolute justice. Laws are not made for men in the abstract—but for real men, placed in precisely determined conditions."[9]

As so often happens, the central tenets of the positivist school reversed, in almost all respects, those of its rival. Where the classicists concentrated upon the crime, the positivists concentrated upon the criminal. Where the classicists saw the offender as rational and responsible, free to choose whether or not to break the law, the positivists saw his behaviour as strongly influenced, if not completely determined, by his innate constitution and immediate environment. Where the classicists insisted that the punishment must be strictly related to the crime, the positivists took the line that it must be related to the offender. Where the classicists saw the sentence primarily as proportionate to the crime already committed, the positivists saw it as a measure for the prevention of future crimes. Where the classicists rejected adaptation of the penalty to the individual personality of the criminal, the positivists insisted upon it. Where the classicists ruled out attempts to reform the lawbreaker, the positivists advocated them. Where the classicists prohibited consideration of whether he threatened future danger, the positivists insisted that his future dangerousness should be the central criterion for deciding whether or not a criminal must be detained for the protection of others. The contrast between the two schools has been well summed up in the aphorism "The classical school exhorts men to study justice, the positivist school exhorts justice to study men."

In their fierce and prolonged confrontation with the classicists, the positivists never managed to take over the field. They got nearest to it in 1919, when Enrico Ferri was invited under royal decree to draft a code for Italy based upon positivist principles.

But when it came to the crunch the code was rejected. One or two South American countries like Cuba adopted measures inspired by it, as did the first criminal code of Soviet Russia, that of 1927. Such concepts as 'dangerousness' fitted in only too well with the need of revolutionaries to eliminate opponents. That, however, did far more harm than good to the reputation of the positivist school.

Though they did not crush their adversaries, the positivists nevertheless succeeded in infiltrating the classicist camp at several vital points. This resulted in a kind of compromise, an amalgam of classical and positivist elements. Their influence was reflected in the proceedings of the International Association of Criminal Law between 1889 and 1914, in those of the Société Générale des Prisons de Paris, even in those of the International Penal and Penitentiary Congresses, composed as they were of cautious and hard-bitten government administrators. It can be traced in the work of Garçon in France, Prins in Belgium, van Hamel in Holland, Von Liszt in Germany. The penal codes of the period, whilst retaining their classical outlines, admitted important positivist modifications. There were radical departures in the treatment of young offenders, longer, even indeterminate, sentences for habitual criminals, extensions of the insanity rules and the idea of diminished responsibility, a new willingness to concede alternatives to imprisonment, like the suspended sentence in Europe, probation in the United States and England. The most harmonious blending of the classical and positivist elements was perhaps that achieved in the Swiss Code of 1937. But the Model Penal Code drawn up by the American Law Institute under the leadership of Herbert Wechsler shows similar characteristics. It has a classical logic and concern for legality in the main grouping of offences in terms of their gravity, and of penalties in terms of corresponding but restrained severity. But this is modified by special positivist provisions fot the young, the mentally abnormal, the dangerously persistent. Alongside the modification of the classical codes came a modification in penal practice, a move away from rigid and impersonal conceptions like solitary confinement towards more diversified and humane attempts to reform offenders rather than merely punish them.

The positivists gave a new and powerful impetus to movements for the rehabilitation of offenders, though they certainly did not initiate them. The desire to reform and restore those who break the law is as old a response to crime as any. It goes back to the old and new testaments, to the medieval church, to Tudor England, to the Rathaus in Amsterdam, to the houses of correction in the Hanseatic cities, to the Quaker penitentiaries, to the avant-garde establishments for children at Mettrai in France and at Elmira in the United States, and to the pioneers of probation. It sprang from philanthropic, religious and humanitarian initiatives, essentially practical and experimental, based on faith rather than systematic theory. But the positivist insistence that treatment must fit personality, that penal policy and welfare policy could interact in the prevention and control of crime, gave a new political significance to this approach, as well as what then seemed a scientific framework and justification.

The towering figure of Sigmund Freud, with his impact on the way people have seen themselves and their experiences, could not fail, in due course, also to influence attitudes to delinquents. Disciples of his who ventured into criminology acknowledged a debt to Lombroso but rejected his classifications and explanations. They used psycho-analytical concepts, particularly to interpret persistent juvenile delinquency. Such phrases as frustration, maladjustment, mental conflicts, anti-social or a-social attitudes, passed into the currency of diagnosis and treatment. With their medical overtones those were significant expressions in themselves. By way of reaction there grew up the suspicion that this whole approach offered elaborate excuses for offenders, implying that they could not help themselves.

Several features of the positivist creed combined to prevent its acceptance as a complete and exclusive system. It took little, if any, account of procedural protections for the suspect or offender. It replaced the familiar idea of criminal responsibility with the somewhat unconvincing concept of "social responsibility," based not on moral guilt but on the requirements of society, and leading not to punishment but to "measures of social defence." In stipulating that the measures taken to deal with offenders should be related to their personalities rather than their crimes, it could be seen as undermining general deterrence. And, in so far as some of its precepts were put into practice, they failed to achieve the impact on recidivism so confidently predicted by its advocates.

The positivist school is today in disarray, indeed it no longer exists as a coherent force, even in its native Italy. Though the theme of individual differences in offenders will always remain important the more far-fetched and sweeping etiological explanations advanced by Lombroso and Ferri were long since discredited. Equally discredited by history has been their assumption that it was no longer necessary to bother about strict legal restraints on the discretion of administrators to detain or to treat offenders. In particular it has become all too clear how dangerous can be the very concepts of "dangerousness," and even of "reformation," as tools of oppression; and how easily measures of "social defence" can be used as weapons of social aggression. Modern research, far from demonstrating the success of attempts to classify and reform offenders, has drawn attention to their failures.

Sociological Approach

The opposition between the classicists and the positivist school initiated by Lombroso should not blind us to what they had in common. Although the positivists took account of social conditions, they thought of them primarily in their impact on individuals. Both schools took it for granted that the starting point was the individual offence and the individual offender. Both hoped to achieve control of crime by way of the criminal law and the penal system. Their battles and their compromises were fought out in the shaping of criminal codes and penal practice. Much more radical is the gulf between them, with their fundamentally individualist outlook, and those who have seen crime as a social product, determined by social conditions, capable of being controlled, if at all, only in social terms.

Adolph Quetelet was the first great interpreter of criminal statistics when they began to appear year by year in the eighteentwenties. In characteristically dramatic style he announced his findings: "We can count in advance how many individuals will soil their hands with the blood of their fellows, how many will be swindlers, how many poisoners, almost as we can number in advance the births and deaths that will take place." And he drew this equally dramatic conclusion: "Society carries within itself, in some sense, the seeds of all the crimes which are going to be committed, together with the facilities necessary for their development."[10]

In a sense all subsequent attempts at social explanations of crime are extensions or embroideries on that doctrine. In societies, as they exist, crime is normal. The facts to be investigated in order to understand crime are social rather than individual, its control must depend upon changes in social conditions. Changing social conditions were indeed producing crime. The state of the cities and of the poorest classes in the heyday of the industrial revolution could be cited as glaring evidence of the overwhelming impact of the immediate environment. To Lombroso's conception of the born criminal the French 'sociological' school, led by Professor Lacassagne and supported by Manouvrier and Tarde, opposed that of the criminal milieu. The swollen cities could be seen as centres of criminal infection and imitation, breeding grounds of crime—or, as Mayhew so graphically described the process in the London slums, training schools for professional criminals. More broadly, the greater prizes offered by a more prosperous society, the lesser risks threatened by a more merciful penal system, the maxims and passions, the cupidity and cunning of a society and business community swayed by greed, all fostered crime. Stability, a fundamental condition of morality, was lacking. The individual could do little against the tide: it was "the great logic of society" which drove him on. It was Lacassagne who coined the now hackneyed phrase "societies have the criminals they deserve."[11]

"Crime is normal because a society without it is utterly impossible" said Durkheim. He went further: "To classify crime amongst the normal phenomena of sociology is not merely to say it is an inevitable, though regrettable, phenomenon, due to the incorrigible wickedness of men; it is to affirm that it is a factor in public health, an integral part of all healthy societies."[12]

The great French sociologist deliberately chose the persistence of crime, so generally regarded as an unmitigated social evil, to illustrate his thesis that we should take nothing for granted about social facts. Since the phenomenon of crime was to be found in all societies at all stages of their development, it must be scientifically regarded as normal: that was a question of fact, not of moral or philosophical judgments. On its usefulness, he had two points to make. First, a society in which there was so great a consensus, so powerful a control, that the law was never broken would be a society so uniform, so rigid that it lacked the flexibility to adapt to a changing world. The presence of çrime thus demonstrated the possibility of moral advance and transformation. In some cases, like that of Socrates, the criminal could be the direct precursor of a new morality. Second, he argued that crime and punishment must be taken together, with the combined effect of reinforcing the major collective values of a society. Where crime weakened them, its punishment re-asserted and strengthened them. This latter theme has been taken up by certain modern criminologists who emphasise the role of criminal law and its enforcement in bringing home to people the specific kinds of behaviour that are most strictly forbidden and, by implication, clarifying the central values of society against which they offend. To put it crudely, punishment of the bad is necessary to assert the good.

The sociological approach to the study of crime, unlike the positivist, emigrated very successfully from Europe to the United States, where it has prospered greatly during the present century, exporting its theories all over the world. Some sociologists would claim to be engaged in a disinterested search for truth, with no ulterior objectives. But criminologists, especially American criminologists, have seldom scrupled to deplore social evils they see as linked with crime, and to advocate changes in social arrangements and penal systems.

They have, nevertheless, shared the positivist failure to produce workable and effective solutions. Moreover, in so far as it tends to portray offenders as hapless victims of their environment, sociology invites the criticism that it is excusing crime. Though many criminologists retain a rather simple optimism about the possibility of engineering social change, their credibility in this respect is being increasingly called into question.

The Socialist Creed

Parallel to the sociological school, but distinguished from it by radical political commitment, was what was frankly known as the socialist school. Its leading exponents were to be found, in the last two decades of the nineteenth century, scattered all round Europe. There were the Italians Filippo Turati and Napoleone Colajanni, the French Paul Lafargue who was a son-in-law of Karl Marx, the Belgian Hector Denis who reinterpreted Quetelet, the German Karl Kautsky who was the leading theoretician of the German Socialist Party, and K.J. Rakowsky who eventually became the first Soviet ambassador to France. Not content with tracing links between crime and various evils that might be regarded as by-products of industrialisation, they insisted that it was bound up with the whole of the capitalist system. It could be eradicated only by the eradication of capitalism. Capitalism deprived the workers of their rights, forced many to live in squalor. Crime was to be seen as the result of their degradation, sometimes the embodiment of their protests.

The socialist school drew its inspiration from Karl Marx, yet neither Marx nor Engels ever developed a systematic socialist theory of crime. For that we have to turn to William Adrian Bonger, a Dutch professor of sociology, a man of great learning and sensitivity and a militant socialist. His book *Economic Conditions and Crime*, published in English in 1916, brought Marxist principles fully to bear on the interpretation of crime in capitalist societies.

Step by step, he implicated the very spirit of capitalism in the genesis of criminality. Capitalism taught men to compete, to accumulate property at the expense of others, to exploit others. Crime involved the robbing or exploitation of other people. Under capitalism the economic mechanism itself, since it set the interests of all in eternal conflict among themselves, made men more egoistic, less compassionate and less concerned about the good opinion of others, and hence more capable of crime.

Capitalism also inevitably produced a class system and that system added characteristic pressures towards crime at each level of society. For the wealthy capitalists there were avarice and ruthless competition. For the small bourgeoisie there was the pre-

cariousness of their position. For the proletariat, above all the
lumpen proletariat, there were the injustices and degradations
under which they lived. It was by these pressures that crime, as a
mass phenomenon, was shaped.

Replace capitalism by communism and all would change. The
spirit of communism was a spirit of co-operation, of altruism, the
very antithesis of crime. The practice of communism would
abolish the corrupting privileges of the rich, the corrupting deprivations of the poor. It would care for those who were weakest in
mind, morality or body, support them and save them from drifting into trouble. In proportion as communism gathered strength,
crime would shrink and wither, starved at its very roots. The
need for criminal laws and penal systems would virtually disappear.

The classicists saw the state as an impartial arbiter, maintaining
a true balance between the rights of the wrongdoer and the
wronged. The positivists saw the state as a paternal authority,
capable of recognising and forestalling danger, of restraining or
restoring those who threatened the common good. The marxists,
on the contrary, challenged the capitalist state itself, seeing it as
no more than a policeman, serving the interests of those powerful
enough to control the means of production. To them criminal
laws and penal systems, like all the other social institutions of
capitalism, were instruments for maintaining the class structure.
The state was not merely an accomplice in this immorality, but
an active participant in that it held down the working classes and
provoked them to crime. Changes in criminal law or in penal
systems were beside the point. It was society itself that must be
revolutionised.

In the days of Marx, even in those of Bonger, the socialist state
was still a dream. It was possible for them, without fear of contrary evidence, to contrast the gross abuses of the capitalist systems under which they lived with the utopian virtues of the
communist system for which they hoped. Now it is nearly sixty
years since their disciples took over Russia, thirty since they
acquired control of East Germany and the other satellites. They
have been able to change the economic basis of the state, but they
have not yet eliminated crime.

Until very recently their reaction to this situation could only
be described as embarrassment inadequately concealed by bluster.

They claimed spectacular reductions in crime but refused any statistics to prove it. Now, from East Germany we are getting admissions that criminality continues, that in some industrial areas it may even be increasing. Of course explanations accompany the admission. Although the economic conditions of life have been changed it may take many decades to change psychological attitudes inherited from the bad old days of capitalism. The state itself may have erred in not giving sufficient weight to education and moral instruction (a curiously Victorian sentiment). The fully communist state has not yet been achieved. But for all this the socialists still twit the capitalist countries with having resigned themselves to crime, whereas they still maintain that socialism will eliminate it in the end.

Meanwhile, they have been obliged, by these very arguments, to go back to the study of the individual offender. Since the whole social environment has been transformed they must turn to individual differences to explain why some people still remain entrapped in the old bourgeois criminality. Since they are activist regimes, calling upon their citizens for ever greater efforts, they again have to admit that not all are equally committed, that something depends, for better or for worse, on individual differences. And they are compelled to assume that those who prove incorrigible, whether as opponents of the regime or as vagrants or other deviants, are either mad or degenerate. Until recently all this would have been regarded as a dangerous deviation from the doctrine that the causes of crime must be sought in economic and social structures. But experience has inexorably brought them back to the individual differences. To quote *Socialist Criminology*, the latest treatise to emerge from East Germany, the individual disposition is to be regarded as "an independent variable amongst the complex causes of crime." That is a significant concession.[13]

Radical Approach

Various groups of "new criminologists," though by no means constituting a single "school" in the traditional sense, reflect a current of opinion, especially amongst the young in the United

States and England. Its main directions can be traced in two
recent collections, both compiled by Ian Taylor, Paul Walton
and Jock Young: *The New Criminology* (1973) and *Critical
Criminology* (1975).

The first to make their mark were American exponents of
what were christened "interactionist" theories. This group re-
jects the study of crime as a separate phenomenon, preferring to
see it as part of the wider field of deviance, which includes not
only all departures from accepted social conventions but much of
what is usually classed as mental illness. Not even stuttering has
escaped their attention.

Since what is regarded as deviant varies from one place to
another, it is emphasized that its definition depends not on the
intrinsic characteristics of the behaviour of people concerned,
but on the interests and prejudices of those with power to enforce their standards. Similarly, whether a particular action or
person is labelled as deviant or as criminal depends on the people
—doctors, teachers, social workers, police, courts—who are entrusted with maintaining these standards. It is a misconception to
assume that a person so labelled is, in himself, sick or wrong, or
necessarily in need of treatment or correction. His behaviour
should be interpreted in terms of what it means to him, rather
than dismissed as wanton or meaningless by conventional standards. Attempts at repression are often self-defeating.

The very measures taken to control or to change an offender
may confirm and extend his deviance. The fact of being stigmatised will alter the way he sees himself, the way others see him
and the way he expects them to react to him. Rejection, or the
anticipation of rejection, may lead him to seek support amongst
similar outcasts, compensating by building up the very deviant
values it was hoped to repress.

Both the distribution of crime, as officially recorded, and the
characteristics of those labelled offenders, largely reflect the
prejudices of those who enforce the law. It is upon the agents of
control, as much as upon those stigmatised as criminals, that research should be focused. Above all it should be concerned with
the interaction between the two.

The more radical criminologists would claim that even this
fails to go to the heart of the problem. It is not, perhaps, unfair to
say that the views of some of them are coloured by strong radical

commitments, with elements of Marxism, Trotskyism, Maoism, anarchism. Many of them would challenge the bulk of existing criminal law and correctional systems, as simply shoring up a social structure that ought to be brought down. They have no time for the conventional approach to the study of crime and its control. In their eyes it starts from a false assumption, in taking for granted the existing criminal law. It compounds its error by using the results and records of law enforcement as the basis for its research into trends and explanations of crime.

The new criminologists would probably concede that classicists were on the right track in wanting to restrict to a minimum the scope of criminal law, the interference with individual liberty. They would certainly endorse the classical concern to control law enforcers, especially the police, together with the suspicion of attempts to reform offenders. The radicals would also argue that classicism did not go nearly far enough, since it took for granted the existing structure of society, the inevitability of the core of established criminal law. For the positivists, in their preoccupation with individual offenders, with defining, segregating or eliminating the dangerous, the new criminologists would have very little time indeed. Even the sociological attempts to measure, analyse, and interpret links between crime and other social phenomena would get short shrift in that they have largely relied upon official statistics.

For the socialist approach the radicals would obviously have some initial sympathy: at least it started from the assumption that the roots of crime lay in the political economy, the very nature and structure of capitalism. But again they would contend it did not go nearly far enough: it still saw the crimes and delinquencies of the downtrodden as maladies and maladjustments, as passive symptoms of the degradation imposed upon them more often than as active protests against it. Nor would the new criminologists stop short at criticism of capitalist attitudes to crime. The treatment of political deviants as criminal in certain socialist countries could be seen as the exact opposite of the kind of society they would advocate.

The radical criminologists have their own utopian vision of a society in which there would be little need of the criminal law. But that would not come about because all citizens were united in a consensus of values. The very idea of such consensus is regarded as a myth fostered by those in power and supported by the assumptions of conventional criminology that individuals can or should be adjusted to some fictional norm. The radical ideal would be a pluralistic society in which many patterns of behaviour, now disapproved or prohibited, would be given full play.

The new criminologists have their points. They have stressed that the criminal law and its enforcement can be used destructively and oppressively by ruling cliques. They have been amongst the foremost in portraying its overextension in ways that can be damaging and self-defeating. They have emphasised the significance of the offender's own view of his crimes and of the measures taken to check them. And the radicals in particular have drawn attention to the dangers of allowing criminology to serve as a mere prop of existing systems. In all these respects they have helped to remove blinkers, to widen our outlooks and attitudes.

Unfortunately they have seemed at times to don blinkers of their own. They have distorted and diminished the impact of their message by their vivid exaggerations. In their anxiety to interpret crime simply as part of the whole spectrum of deviance they have blurred important distinctions, resorted to misleading analogies. The radicals have felt bound not merely to question but to contradict all that has gone before. They have talked as though conventional criminologists were unaware of the limitations of criminal statistics, and as though the statistics were rendered completely valueless by those limitations. They have overstated the heterogeneity of social values, ignoring the large measure of consensus, even amongst the oppressed, in condemning the theft and violence that make up the bulk of traditional crime. They have postulated an irreconcilable opposition between "law and order" on one hand, "human rights" on the other, blinding themselves to the genuine functions of law, police and courts in safeguarding people's rights. And they have indulged in exaggerated hopes of human nature, been overoptimistic about what society will tolerate, either now or in the future.

Radical criminology has not been alone in going to extremes. All new movements, in their first flush of enthusiasm and their determination to achieve the maximum impact, begin by distorting or denying the achievements, reversing the values and emphases, of their predecessors. All attitudes towards crime have their periods of challenge and eclipse. Yet they tend to persist, sometimes reappearing in fresh guises, under new names. Often they have to learn to co-exist within a single system. They are perforce modified by their own experience, by mutual influences and antagonisms, and by the changing climate of the times.

Attitudes to the causes and control of crime are not the same thing as explanations. Yet they powerfully influence the directions in which explanations are sought, the ways in which findings are interpreted.

CHAPTER
4
The Search for Causes

It is natural for people to ask questions about crime, to expect straightforward explanations. The broadest questions come from those who think of it as a single entity. Why is there so much of it? Why is it rising? Why has it occasionally gone down? Why does criminality differ from one neighbourhood or country to another? Others are more concerned to probe specific kinds of crime. Why are there more robberies? Why is vandalism more prevalent? And there are those who want to know why certain people turn readily to crime whereas others do not, or why similar social pressures produce so wide a range of responses. Going further, they may ask whether we cannot identify potential criminals and thus predict and prevent criminal behaviour.

The systematic search for explanations of crime by way of empirical evidence has taken two forms. One is primarily sociological, preoccupied mainly with the trends and distribution of crime. The other is primarily psychological or psychiatric, preoccupied with the question of how and why particular individuals become criminal. Of course it is no more possible in criminology than in life to keep society and the individual in separate, hard and fast categories: they overlap and interact perpetually and at many points. Social-psychology can in some ways be seen as an attempt to comprehend both. But each has its distinctive questions and ways of trying to answer them.

The Social Setting

The kinds of comparison used in efforts to tease out social factors linked with crime can likewise be divided into various groups. They include comparisons over time, asking such questions as "What changes in social or economic conditions are linked with changes in the level and forms of criminality?", and also comparisons between groups or places, asking such questions as "What difference in their social conditions seem to be linked with differences in criminality?" Some envisage social factors, such as poverty or wealth, as directly impelling or tempting people to crime. Others see them rather as weakening the social controls that might inhibit it.

It is remarkable that so little systematic research has been directed to the occasional instances in which the rise in crime has slackened, even been temporarily reversed, or to groups amongst whom crime has remained comparatively rare. The fear, fascination, or frustration provoked by criminal activity often seems a necessary spur to arouse the interest of researchers and bring in the funds to support them. Similarly, the great bulk of investigation has concentrated upon what are seen as clear social evils, such as poverty, slums, lack of education and opportunity. Though from time to time criminologists have inveighed against the crimes of the wealthy, it is only comparatively recently that systematic research has invaded this preserve.

Whereas the seeds of almost all modern sociological interpretations of crime were sown in nineteenth and early twentieth-century Europe by such criminologists as Tarde, Durkheim, Bonger, it is in the United States that they have been developed. The hypotheses put forward, the research embarked upon there in the course of the present century have been enormous in quantity, but very variable in quality. Only a small fraction of all that has been published is worth serious consideration, and no more than a few major trends can usefully be examined here.

There are at least five good reasons why the United States has so far produced most of the major attempts to explain crime in social terms. It has been in the States that affluence has reached its most spectacular heights. It has been there that the combination

of crime with an advanced civilisation has assumed its most
threatening proportions. It has been there that people from all
over the world have been brought together to be welded into the
most powerful nation on earth. It has been there that, since the
nineteen twenties, empirical studies of many kinds of social phenomena have been undertaken and supported on a lavish scale. It
has been there that confidence has flourished in the ability of a
free society to solve its problems. With so much achieved in
other ways, it seemed that the understanding and control of
crime should surely come well within the range of competent
social engineering.

Adapted, modified and enlarged, the criminological interpretations developed in the United States have achieved world-wide
currency, the more so as the rise in crime has come to be recognised as an obdurate problem virtually everywhere. Their adoption resulted partly from American prestige and presence; partly
from a dearth of bold contemporary explanations elsewhere;
partly from the shortage, in other countries, of resources for
large-scale empirical research; partly from the expectation that
the problems of the United States would soon be those of
Europe; partly from the realisation that the developing world, as
it moves into industrialisation, is already exhibiting features of
similar malaise. Each of the major themes of American criminology is now a commonplace in any gathering, academic, administrative or political, for the discussion of crime and its control.
Even Soviet Russia has found it desirable to have translations of
some of the leading texts. She may want them only to refute
them, or to exploit them as evidence of the rottenness of capitalism, but she cannot ignore them. And in Scandinavia, England,
the Commonwealth and elsewhere, they have been put to use as
the jumping off ground for research: yet again to be amended,
contradicted or extended to fit different conditions.

Of all the social factors linked with crime, economic conditions
have attracted the most attention. Attempts to relate crime to
poverty, in particular, run right through the history of criminology, and indeed go back far beyond that. There are references in
the writings of Xenophon, Plato, Aristotle, Virgil and Horace, of
moral and religious philosophers such as Thomas Aquinas and
Thomas More, and of the French philosophers of the enlightenment, as well as of many later social enquirers.

It seems at first sight so obvious, so simple, that men steal because they are poor. The prevalence of poverty or prosperity stand out so much more vividly and distinctly than many other factors in social life. One need not be a marxist, or even a social-ist, to appreciate their massive influence. And because they ap-pear to represent relatively concrete phenomena, capable of precise measurement, they offer tempting material for investiga-tion. Moreover, the fact that they can change so sharply and abruptly adds to their attraction for comparative research.

As usual, however, the apparent simplicity vanishes as soon as we begin to investigate them. First, when we are talking of pov-erty, what do we mean? Few people in developed countries today face absolute destitution and are forced to steal as the only way of feeding their families. It is true that in earlier eras of history, and in the poorest parts of the world today, sheer destitution was, and still must be, a direct and compelling factor in crime. But when it comes to the kind of chronic peasant poverty that is almost the rule for vast numbers of mankind, we are looking at something quite different. Quetelet found that the poor rural areas of Europe were usually the most honest. It was the poor of the cities, living cheek by jowl with great wealth, who were under the greatest provocation and temptation to crime. The key there lay, he believed, not in absolute poverty but in relative poverty.

The theme of relative poverty was taken up, both in Europe and the United States, in studies of the effects of depressions on levels of crime. That Nestor of sociological investigators in nine-teenth century Bavaria, Georg von Mayer, reached the conclu-sion that for every additional gulden on the price of corn there were fifty more thefts. However clear the apparent connection in a mid-European peasant community, no such definite link be-tween an industrial depression and theft had been found amongst English factory workers earlier the same century: in at least one area they had been discovered to remain doggedly honest in the face of adversity.

Certain conclusions, however, emerge from later and more sophisticated investigations. A sudden drop in prosperity creates a wider and stronger pressure towards crime than accustomed poverty. Even so, impact tends to be delayed, as though the moral momentum took time to adjust to the economic. Moreover the rise in crime is less marked than the decline in prosperity. Nor do all kinds of lawbreaking move in the same direction. Whereas larceny tends to expand during a depression, sexual and violent offences have been known to contract. Even amongst crimes of theft there are differences: the picking of pockets, for example, would go down in a depression, when they were more likely to be empty, up in prosperity when they were more likely to be full. Fraud and embezzlement flourished under both conditions: in poverty there is the temptation to keep afloat at all costs, in wealth the pull of greed and opportunity. Finally, and most unfortunately, there is scant evidence that a return to prosperity following a depression brings with it a return to previous standards of honesty.

The riddle facing the prosperous countries in the decades following the second world war has been the steep and prolonged rise of crime under conditions of growing affluence, full employment, and widening provision for education and social welfare.

The clearest and bluntest version of the theory that rising prosperity entails rising crime had long ago been advanced by a nineteenth-century Italian called Filippo Poletti. He postulated that in any society the amount of criminal activity must be related to the amount of honest activity, as a kind of inevitable waste or by-product. If expansion and progress result in more transactions, more business and social intercourse, more property everywhere, they will also result in more opportunities for crime and therefore in more crime. On this basis he argued that the true criminality of a nation should be measured not just by the number of crimes in relation to population but by the number of crimes in relation to lawful economic and social transactions. So long as crimes were not multiplying faster than these you could not really say that society was becoming more criminal.

In the bad old days in Cornwall, when ships were wrecked on the coast, whole villages became criminal at one fell swoop. They flocked down to gather up the goods. They even put out false lights to ensure a plentiful supply of shipwrecks. Nowadays we have no need of that. A prodigal economy showers property everywhere: in shops, offices, factories, in cars and on the streets. Is it any wonder that people pick it up? The very mechanics of life widen the opportunities, reduce the inhibitions. The big business or state-controlled firm is remote and impersonal; there is less compunction about cheating it. The cheque, the credit card, the computer, have opened up vast new opportunities for deceit. Far more is being regularly spirited away by computer frauds than by the biggest train or bank robberies. The villagers could hope to enjoy with impunity their ill-gotten spoils, since their neighbours were equally implicated and unlikely to betray them. That still applies in some situations. We have now to add the opportunities of undisturbed enjoyment of the fruits of crime opened by modern mobility, both geographical and social. Abundant temptations to transgress are reinforced by abundant chances of permanent impunity.

Such arguments have never been popular. Poletti's ideas were met with violent resentment from all but a handful of his contemporaries. No society likes to be told that the achievements it most prizes are achieved at the cost of encouraging crime. And many regard it as an unjustified slur upon human nature to suggest that greater opportunities for getting away with crime will almost always mean that more crime is committed. Yet psychological research, as well as experience, offers plenty of evidence that this can be the case.

The more modern American interpretations of relations between economic factors and crime have been more sophisticated, whether with regard to poverty or to plenty. They have envisaged the impact of economic conditions as indirect, affecting crime by weakening social controls, or people's respect for them.

It no longer seemed enough, for example, to explain high rates of crime in certain city slums as due to squalor and poverty in proximity to plenty. Since not all such areas were criminal, what was it that distinguished those that were? It was suggested that the answer lay in a breakdown or conflict in social organisation. Successive waves of immigrants were moving into certain city centres, where they could find the cheapest housing to begin their new life. Later, those who prospered and were successful moved on, amongst them the potential community leaders. These areas were thus left to the weak and the failures, those least able to maintain values, or initiate action. The result was a breakdown of informal social controls, pockets of social disorganisation in which delinquency could develop virtually unchecked and be passed on from one generation to the next. If, into the bargain, such areas were taken over by criminal rackets, two competing systems of values and controls could develop side by side: that of the generally law-abiding but incompetent poor and that of the criminals, whose conspicuous success in material terms was a challenge, as well as an affront, to conventional standards.

The concept of anomie has been developed in a more general way as the theme of diminishing respect for authority. The origin of this strange term was an English word, which first appeared in 1591, with the meaning of disregard for law. Emile Durkheim adopted it and applied it to fiercely rising commercial competition, which had broken through the traditional rules and restraints. He suggested that the limits formerly imposed by society upon the aspirations of individuals in different classes of the community had been swept away as a result of the overriding importance attached to industrial advancement. The theme was later taken up in the United States and applied not merely to industry but to the basic social and political outlook of the country. In addition to the primary importance ascribed to material progress there was the emphasis on the democratic ideal of equal opportunity for all. But it was impossible for everyone to succeed in the race for riches, and those who failed to achieve success in these terms found themselves branded as failures in moral as well as material terms. Yet the race was not a fair one, especially for those who started out with financial, social and educational handicaps. Society put before people the one goal of getting rich, but many, especially in the poorer classes, could see no hope of arriving there if they stuck to the rules for doing it legally. Few of them, for example, could get the good jobs, the social contacts, the initial finance, to launch them on the road to success. Inevitably they were tempted to lose confidence in the social structure and regulations that promised them achievement yet doomed them to failure.

It was not claimed that all who were faced with this basic conflict would become criminal. A few would fight their way through to legitimate success in spite of the obstacles. A few would become ultra-conformist, meticulously law-abiding without hope of reward. A great many would shrug their shoulders and accept the situation. Some would sink into depression and retreat, rejecting the very idea of getting on in life, perhaps resorting to alcohol or other drugs. But a few would respond by outright rebellion, attempting to overthrow the existing system, to replace it by something quite different. And some would, quite cynically, accept the conventional goal of success but adopt crime as a short-cut to achieving it.

Definitions and criteria of anomie have been vague, sometimes conflicting, sometimes circular. That, I am sure, would be accepted by Professor Robert Merton, who has done more than anyone to develop the theory. Attempts to test its impact on criminality have not shown a simple relationship. For instance, the evidence suggests that it is the "moderately anomic" and the "moderately conformist" rather than the most alienated who are the most criminal. Moreover people can move from one response to another, at one stage perhaps rebellious, at another conformist.

The role of culture-conflict in presenting people with competing rules and values has already been touched upon in connection with crime and immigrants. It has, however, a much wider significance in the modern world. There is the speed of change, the values and beliefs of one generation, even one decade, being challenged or abandoned by those of the next. There has been the opening up of the whole world by quick and easy travel, above all by press, radio and television. Contrasting beliefs and ways of life are seen side by side. A generation or two ago it was possible to grow up in the firm conviction that the way things were run in your own country was beyond question, virtually immutable. Now everything lies open to comparison and to dispute. That includes the criminal law, the system it supports and the duty of the citizen to obey it.

Yet, in spite of all these pressures towards crime, in spite of all these challenges to social controls and accepted values, the great majority of people do not become persistent criminals. There are many slum districts which are not delinquency areas, many disadvantaged groups which do not respond by violence or lawbreaking. Why is it, then, that one individual or group reacts to deprivation or opportunity, culture conflict or anomie, by resorting to crime, others do not? Given these underlying factors, affecting the whole society or large segments of it, why the differences in response?

Edwin Sutherland, author of perhaps the most widely read of all textbooks on criminology, propounded his "theory of differential association" as an answer to such questions. He envisaged it as a comprehensive theory, which would always and decisively differentiate those who committed crime from those who did not.

In its original and crudest form it simply stated that the chances of becoming criminal depended upon the balance of influences for and against law-breaking which a person encountered in the course of his life. If the majority of attitudes he met were criminal, he would become criminal. If they were anti-criminal, he would become law abiding. It was simply a matter of learning like anything else. As a result of objections from critics taking a rather subtler view of the processes of learning, Sutherland later conceded that it was not merely a matter of adding up the number of contacts: weight must also be given to how frequently they occurred, their duration, how early in life they were encountered, and the prestige, in the eyes of the receiver, of the people or groups conveying them. What he would never concede was that people might become criminal without learning from direct contact with others the motives, attitudes, rationalisations and techniques belonging to crime.

Sutherland contended that his theory could explain both individual differences in response to similar social pressures and differences between groups. Thus the persistence over generations of lawbreaking in delinquency areas, or of violence amongst black adolescents, could be explained by the virtual inevitability that youths would learn such attitudes from their contemporaries. The greater likelihood of crime amongst boys than amongst girls could be explained by their greater freedom to make associations and pick up delinquent ideas on the streets. Differences in individual disposition, if they came into it at all, did so only indirectly. Since crime was not the only possible response to mental conflicts, any connection between emotional disturbance and delinquency must be explained by the tendency of the maladjusted to stray away from home, thus running more risk of contact with criminal patterns outside and escaping the law abiding influences of their parents. Indeed, in a delinquency area, the normally sociable and active boy, wanting to be out and about, could be more at risk than the neurotic or retiring.

Sutherland left it on record that his theory had been sparked off by a conversation with a professional criminal. It was this that convinced him that, in order to adopt crime as a way of life, it was essential to learn both the techniques and attitudes of established criminals, much as one learns the techniques and attitudes of a profession. But the attempt to generalise from this to the whole field of criminality has hardly stood up either to theoretical criticism or to empirical testing. A study of embezzlers by Professor Donald Cressey led to the conclusions for example, that the techniques did not need to be learned from other embezzlers, they were simply the techniques picked up in the course of the offender's normal job; the attitudes which made it possible for a man to embark on the offence were not derived from sources that could easily be identified or measured. Moreover, it had also to be conceded that a person might be driven to crime by psychological or physical deficiencies that prevented him from meeting his needs in more acceptable ways.

Given all these concessions we are left with little more than the conclusion that some criminal attitudes and techniques, like many other attitudes and activities, are learned from others. The part of the theory that has been neglected is that law-abiding attitudes need to be learned from others. It could well be argued that it is inadequacies in this kind of learning, rather than active learning to transgress, that most often tips the balance.

Interpretations of the trends and distributions of crime in terms of other kinds of social pathology have their value. In most countries there are sections of the population living under the pressures of poverty, neglect and social injustice, tempted to crime, exploited by criminals. But it cannot be pretended either that such conditions always produce crime or that crime is confined to people living under them. Ramsey Clark, a former Attorney General of the United States, devoted a considerable section of his book *Crime in America* to the squalor and the crime of the slums, commenting that "children living in places where people have no rights that they are capable of enforcing will rarely have a regard for the rights of others." But he had also to acknowledge that white-collar crime was usually the work of respected and successful people: "The trusted prove untrustworthy; the advantaged, dishonest."[14]

Interpretations that link crime with material and social progress are more realistic in taking account of white-collar crime, together with the vast dark figure of offences committed by comparatively prosperous citizens in the course of their normal occupations. The extension of physical and social mobility and of contacts with different cultures and values; material abundance; freedom of enterprise and choice; ambition and the will to achieve: all these can be seen as contributing their quota to crime at every level of society.

"If the tree of crime, with all its roots and rootlets, could ever be torn out of our society, it would leave a vast abyss," said Gabriel Tarde, almost a hundred years ago. Today we have even more reason to acknowledge the truth of that.

The Individual Potential

I was introduced early to the individual propensity to crime. Sitting in a Viennese café on my way to study in Rome, I looked up to find a stranger staring at me across the table. I stared back. Demanding "Why do you persecute me?" he hit out. When I put my arm across my chest to protect myself he produced a revolver and took a shot at me, penetrating my elbow. They said he suffered delusions and put him into a mental hospital. When I visited him some years later he was still there.

Why put the blame for rising crime upon society? Every criminal act is the work of one or more individuals, the fruit of individual characteristics and attitudes. Under any social conditions, however good, there are some who transgress. Under any social conditions, however bad, there are some who do not. If we are trying to understand trends of crime in society we have to look at social changes and differences. But if we are trying to understand what kinds of people commit crime, and why, we have also to look at individual differences.

However, it is possible to oversimplify even in that. Every individual is the product of the environment in which he is born and reared as well as of the constitution he has inherited through his parents. The two constantly react on each other from the first beginnings of life. You could say that what a person inherits is potential, say for high intelligence or for schizophrenia. Whether the potential is fulfilled, for good or ill, depends largely on his environment. Certain kinds of mental illness have, however, tended to become closely associated in people's minds with certain kinds of crime. For example, schizophrenics, particularly those suffering from paranoia, depressives, the mentally sub-normal and epileptics, have all been regarded with suspicion as especially prone to violence or even murder, perhaps also to a range of lesser crimes. How far are such suspicions justified? And what of the epithet psychopathic, attached to such a variety of persistent offenders? Does it represent some personal abnormality, capable to specific diagnosis and providing a constitutional explanation of their criminality? Or is it no more than a label reflecting the view that they are a social menace or social nuisance?

There is indeed the occasional extreme case, where murder seems to stem direct from madness. There was Daniel M'Naghten who, without apparent provocation, shot the Private Secretary of Sir Robert Peel in Whitehall in the eighteen-forties. The discovery that he had acted under insane delusions of persecution led to the framing of the M'Naghten Rules on the criteria of insanity to be used in criminal courts. There are the men who find themselves in Broadmoor, having suddenly murdered their wives, after years of grumbling jealousy in what otherwise seemed normal lives. There are the Jack the Ripper types of murder, cold blooded atrocities which may stem from schizophrenic fantasies in which women appear as sexually polluting witches. And there are the more frequent and tragic occasions when a parent, a husband or wife, kills a whole family under the delusion that they, or the world at large, are doomed to some yet more terrible fate. Or again there is the occasional instance where someone who has started life in quite promising circumstances mysteriously begins to deteriorate in personality and behaviour, both at work and in private life, until suddenly he commits a savage assault or murder. The source of the trouble may be some accident or disease affecting the brain. Such stories linger in the imagination. The more terrible they are the deeper the impression they create, the more lasting the fears they produce.

This depth and persistence of impact can give a totally false impression of the danger of violent crime from the mentally ill. There are the occasional instances, especially in paranoia, depression or brain damage, where the illness is not recognised in time to forestall tragedy. But severe mental illness is usually too incapacitating to make crime, or any other social activity, feasible. Moreover it is reasonably easily diagnosed, the patient under care and a measure of restraint during periods when he might be a menace to himself or others. The proportion amongst psychotics discharged from hospital who subsequently kill or commit serious assaults is little higher than the proportion amongst the population at large. Where they are a danger it is to their families rather than to people outside.

With minor crimes it is a different matter. The weird delusions and distortions that afflict schizophrenics make it difficult for them to cope with ordinary life. Even after treatment they may remain apathetic, aloof, ineffectual. They are common amongst social derelicts, drifting into destitution, vagrancy, stealing food or clothes when they get the chance. There are milder forms of pathological depression, linked with far milder forms of crime, like that of the women, occasionally found amongst shoplifters, whose distress has preceded their thieving and is not merely the result of being caught. Only rarely, however, does neurosis manifest itself in crime. Its more usual effect is to act as a barrier against it, making people anxious, inhibited, overconformist rather than rebellious. Among the many epileptics, only a small minority are suffering from brain injury, or manifest their illness in violence or any other kind of crime.

Psychotic cases are in the minority even amongst murderers. And when it comes to offences as common as petty theft and shoplifting, they are a very tiny minority indeed.

Of people classified as mentally subnormal, well below the average in intelligence, much the same can be said. The really severe cases are easily diagnosed and live out their lives under close care, whether from their families or institutions. They have seldom either the capacity or the liberty to indulge in crime. Those whose congenital handicap is less severe may never be labelled as subnormal as long as they have families or friends able to protect and support them. The great majority, indeed lead placid sheltered lives, no threat to others and with no predeliction for crime. Only a minority, coming from backgrounds and homes which aggravate their disabilities or lay them open to criminal associations, are led into repeated, though unremarkable, delinquencies. Although, like the chronic schizophrenics, these are over-represented amongst persistent petty criminals, they still contribute only a small share to crime as a whole. They are a nagging problem to the police, the courts, the welfare and penal systems rather than a real danger to the community.

Thus the segments of crime and criminality to which mental illness or deficiency offer us keys are now and then tragic, even appalling, much more often just tiresome and obdurate. But they are relatively small. If we want to find broader answers as to why some people become criminal whilst others do not, we must look at the range of so-called normality. The idea of identifying certain traits, apart from criminality itself, which will reliably distinguish the true criminal from the rest of us, has a perennial fascination. Three particular themes have been singled out: that criminals are people of inferior intelligence; that they are subject to some kind of chemical imbalance; or that they are different in physique, and therefore temperament, from the more law-abiding.

The idea that criminals were mostly dullards or illiterates flourished in the nineteenth century. It lost a good deal of credibility when it came to be recognised that studies based on people arrested, let alone detained in penal institutions, were heavily biased in terms of intelligence, indeed that the sheer fact of prolonged detention could contribute to the impression of dullness. With more sophisticated modern investigations, comparing people at liberty and of similar background and class, the differences are far less startling than was once believed. The young delinquent, for example, may often be well behind other boys in schooling but he is only a little way behind them in intelligence.

There used to be much speculation about the influence of glands on behaviour. For instance, might not some peculiarity of the endocrine glands produce the kind of person liable to commit crimes of impulsive violence? Little, if anything, has come of all that: though we have learned since then to measure much more accurately the levels of hormones in the blood, no convincing evidence has emerged of consistent differences between criminals and the rest. Recalling the marked differences in the criminality of men and women, we might think sex hormones had something to do with it. It has been demonstrated that a submissive hen can be converted into an aggressive cock-like creature by dosing her with testoterone. But the same treatment given to women, though it may make them more sexually responsive, does not alter their temperament. To administer female hormones to adult men, to dose them with a chemical which reduces the effect of male hormones, can reduce libido and with it the temptation to sexual offences. But it does not affect their aggressive feelings, or any tendencies they have to violence or other crime.

In modern terms the question of chemical balance has arisen mainly in connection with mood changes, including those it is fashionable to call 'altered states of consciousness.' People vary, for example, in their response to a drop in the level of blood sugar, caused by going too long without food or, more dramatically, by an overdose of insulin. It often causes irritability, impulsiveness, with some people recklessness. In rare cases there may be outright violence, as when a man who was particularly susceptible to this phenomenon murdered his mother at such a time. Much more common are chemical changes induced by the abuse of certain drugs. An overdose of amphetamine, such as used to be taken, for example, as French blues or purple hearts, can cause pugnacity or a dangerous keyed-up state that can lead to serious violence. This would end in crime much more often if it were not so easily recognised and medically controlled.

The links between crime and alcohol are of long-standing and well known. It is an important contributing factor in crimes of violence. As with all other drugs, the vulnerability both to intoxication and to long-term addiction varies greatly between individuals. But the liability to fall foul of the criminal law, whether for being drunk and disorderly, or for assault, theft or damage whilst drunk, depends very largely indeed on the situation in which the trouble occurs and the social standing of the person concerned.

The third and very popular way of trying to sort out the black sheep of the human race depends on the assumption that physical shape and appearance is an indication of temperament, and that temperament, in turn, is the key to delinquency. The older theories gave most weight to the shape of the head, the face, the features. More modern investigators have concerned themselves with body-build and have frequently come up with the finding that, in comparison with non-delinquents of similar age and class, the athletic, muscular category is somewhat over-represented amongst delinquents. You can see that the beefy type might be best fitted for rough criminal exploits and therefore more inclined to take to that way of life. But a more subtle connection has been suggested. Tests indicate that there is indeed some correlation between temperamental traits and physique, and the muscular type is said to be more extravert than the average. If we also accept the theory that extraverts are more likely than introverts to become involved in crime, we have a longer series of links leading from physique to delinquency. But it is all very tenuous. The differences in delinquency associated with body types are minute. Real people seldom fit neatly into a single type, whether of physique or personality. The vigorous, athletic type is as fitted to become a good sportsman as a good delinquent. Which he does become depends largely on the circumstances of his life. What would happen to all these findings if we could add the hidden delinquencies of white-collar criminals?

When we come down to the one temperamental characteristic that does seem to have a strong association with delinquency, we are getting very close to delinquency itself. This is the degree of aggressiveness. If we compare young children, before any official career of offences begins, it is possible to pick out, by personality tests and observation, some who are noticeably more assertive, daring, prone to fight, quick to anger, resistant to adults and to discipline. On average these are the boys likeliest to become involved in later delinquency and crime. This is the nearest we seem able to get towards finding a 'criminal personality.' And it is very far from providing an infallible test.

At the extreme of this range of awkward characters is the amorphous collection of people popularly lumped together as psychopaths. They are very troublesome individuals, unrestrained in their behaviour, callous, impulsive, unresponsive to ordinary deterrents. They are not mad but they are not normal either. Some are given to uncontrolled violence. Some just seem unable to face the normal responsibilities and irritations of life: they cannot get or stay in jobs, they fail to marry or drift away from their families if they do. Unable to cope with minimum social expectations, they drift into petty crime, like some chronic mental invalids, as the easiest way of satisfying immediate needs.

The term 'psychopathic' to describe such people was first coined in Germany in 1888, but the idea that they suffered some mysterious moral insanity or moral imbecility goes back at least to the early nineteenth century. Their resistance to deterrence and to attempts to reform them fuelled protracted debates in European Penal and Penitentiary Congresses around the turn of the century: was there or was there not such a being as a truly incorrigible offender? The conception of the psychopath enjoyed its greatest vogue, however, in the years between the two world wars, preoccupying criminal lawyers, psychiatrists, penologists, cropping up all over Europe and the United States, making conquests even amongst the calm and cautious Scots. One German Professor of Psychiatry, having published a book of several hundred pages on 'diminished responsibility,' had the misfortune to be called in as an expert to diagnose Adolf Hitler when he fell foul of the law in the early stages of his career. He described him as a psychopath and was lucky indeed to lose no more than his job when Hitler came to power a few years later.

The psychopathic personality has been debated, dissected, classified. We have been told that there are aggressive psychopaths, sex-psychopaths, inadequate psychopaths, passive psychopaths, each group manifesting its basic cussedness in a different way. There have been estimates of the proportion of such people amongst offenders coming before the courts, amongst persistent offenders, and amongst prisoners. Figures of anything from ten to ninety per cent have been bandied about, depending on how psychopaths are defined and on the group being studied. Take a population of recidivist prisoners, and a broad definition of the psychopath and you can claim to have 'explained' the greater part of them in terms of psychopathy of one kind or another. But have you really explained anything at all?

There are as many different definitions of psychopathy as there are psychiatrists willing to define it, and a growing reluctance to define it at all. In Germany it is still in common use. In the Scandinavian countries it has largely been abandoned. In Denmark it is said that psychologists and social workers dislike it whereas psychiatrists still find it useful. In the States, where in some parts it had been so widely accepted as to figure in legislation, there are tendencies now to react against it and all it implies. One English expert put it quite bluntly a few years ago: "A psychopath is somebody we don't understand and can't treat."

In spite of intensive investigations no effective criteria have been discovered to differentiate a psychopath from other people. Medical research has uncovered certain characteristics which, it has been suggested, are unusually prevalent amongst men in penal institutions who have been classed as severely psychopathic. A substantial proportion of them, though by no means all, have abnormal brain waves, revealed by electroencephalogram. A much smaller proportion have an extra male chromosome. But a great many people who display the behaviour ascribed to psychopaths have neither of these characteristics. There are certainly large numbers of noncriminals with abnormal brain waves. The double male chromosome is extremely rare anyway, but when it has been detected in institutions it has been found more frequent in hospitals for the subnormal than in prisons. We have yet to discover whether its incidence is really any higher amongst alleged psychopaths than amongst the population at large. Certainly neither this nor abnormality in brain waves could be used as a diagnostic test, enabling a doctor to say that a particular person was, or was not, a psychopath.

Another possibility still under exploration is that psychopaths are distinguished by a slow physical response to stress. The reasoning is that this makes them unresponsive to the kinds of discipline that teach most of us to accept social restraints. As a result they do not learn self-control and react explosively to frustration. Findings are, as yet, too uncertain and confused to provide convincing evidence that this is indeed a genuine constitutional characteristic of the people classed as psychopaths. And there are plenty of people who resort to violence in fits of explosive rage, especially within their families, without showing any of the other characteristics, like chronic callousness or irresponsibility, ascribed to the psychopathic personality. Even if some distinctive constitutional factor is eventually discovered, the likelihood is that, in this as in so many other characteristics alleged to predispose people to crime, it is a matter of degree, its outcome heavily dependent upon the circumstances of upbringing and environment. Though there may be a congenital disposition towards violence, it is by no means inevitable that it will express itself in crime.

Over the whole field of mental illness, subnormality, so-called abnormality, as over the field of crime, there are two warring schools, almost two ideologies. One stresses the reality and role of individual peculiarities. The other stresses the overwhelming importance of social pressures in bringing out, even creating and defining, such peculiarities.

Thus, on the first view, schizophrenia or depression would be seen primarily as illnesses that run in families, fruit of a constitutional predisposition, linked with definite biochemical differences. On the second view, the burden of responsibility would be laid on families who chose to make one of their members a scapegoat, on the role of emotional stress in precipitating attacks. In several of the more extreme forms of mental subnormality, such as mongolism, the question has been settled by the identification of specific physical causes. It used to be fashionable, also, to ascribe the common, less severe cases to heredity. Family trees were drawn up in America to demonstrate that succeeding generations reproduced features of mental backwardness, social inadequacy and crime. Then it was pointed out that the likenesses could be explained by upbringing and example just as well as by inheritance, and we got the view that it is this social factor that really decides whether a person becomes, or is classified as, mentally subnormal.

Yet again, in debates about psychopathy, there are those who devote their energies to the search for evidence of physical abnormality. And there are others who are satisfied that the main explanation, if not the whole, can be found in such experiences as parental rejection, brutal or erratic discipline, growing up in the deadening atmosphere of institutions. Here, as elsewhere, the interaction of individual and environment is apparent. Quite apart from the risk that the difficult child will have difficult parents, the awkward baby can provoke his mother and father, setting up a vicious circle of retaliation and rejection which extends later into relations with school, police and other authorities, perhaps making removal from home inevitable, care by fosterparents outside an institution impracticable, delinquency very likely indeed.

The dispute about the proportionate roles of nature and nurture is very old. The idea of the born criminal, or at least the inheritance of a predisposition to crime, was once extremely popular. Now, with our democratic desire to believe that everybody starts equal, we have swung towards the other extreme, attributing the lion's share of responsibility to family and social environment. But the stubborn fact remains that, from the beginning, individuals react differently to very similar circumstances. A study of Danish children adopted at birth demonstrated that their criminal records were more likely to resemble those of their natural fathers than those of their adopters, quite striking evidence that some genetic factor was at work and that environment and upbringing are not all-powerful.

Why So Little Impact?

Why should it be that a century of theorising and research should have made little or no apparent impact either upon the trends of crime in society or upon our ability to modify criminal tendencies in individuals?

First there are the obstacles to 'explanation'. It has become increasingly clear that straightforward comprehensive answers do not exist except in terms so broad as to be truisms. To say that crime is the result of sin or greed, frustration, aggression or oppression, is no more than a first step. These are very general human conditions, but they produce many other reactions besides crime. Seldom, if ever, does a single explanation provide more than a partial truth. It is not a matter of "either-or," or even of "both," but of the convergence of many factors which constantly affect each other. Moreover, crime covers a vast range of behaviour which can no more be covered by a single explanation than can disease. And even if the search for explanations is narrowed to particular categories of crime, it is still very unlikely to lead to a single 'cause', since very different meanings and motivations can lie behind what at first seems a precise legal definition. That applies to all forms of crime, from murder to petty theft.

If we retreat from talk of 'explanations' and 'causes' in such general terms to a more modest search for the factors contributing to crime, we still face complications. Take three phenomena that have very commonly been regarded as conducive to crime: poverty, broken homes, maladjustment or inadequacy. These mean very different things and have very different impacts in different parts of the world, at different periods of time. Moreover, the more we try to seek out and isolate factors leading to crime the more we find combinations of interacting pressures rather than a single decisive influence.

Again, though criminological research claims to be inductive not deductive, based on empirical observation rather than speculation, it virtually lacks the possibility of experimenting in the full sense of the word. Crime and punishment cannot be isolated from other influences, transported into a different environment, manipulated like the non-human subjects of scientific experiments. Factors found to be associated with crime do not necessarily cause it. That a person suffering from some mental disability commits an offence does not inevitably mean that he did so as a result of his disability. Varieties of social pathology found in connection with crime may spring from the same underlying sources as the crime itself.

Second, even where links between criminality and other social or personal factors seem indisputable, it by no means follows that we are able to prevent or change them.

Neither exploration of the social evils linked with crime, nor acknowledgment of its ties with what has been counted as progress in social and economic development, offer a royal road to controlling it. As I said ten years ago, in a report on *The Need for Criminology* presented to the Bar Association of the City of New York and the Ford Foundation, there is certainly a case for attempting to ameliorate social conditions and improve the lot of the poorest, huge as are the resources needed to do so. But that case rests upon justice, humanity, political wisdom. There is no firm evidence, certainly no guarantee, that it will result in a reduction in crime. And what chance can there be that liberal nations will sacrifice the hard-won fruits of material progress, the world-wide mobility of people and ideas which enlightens and enriches as much as it challenges, simply in the hope of reducing crime? We accept, with apparent equanimity, that more cars mean more road casualties, more smoking means more lung cancer, more drinking means more alcoholism. All these tragedies we are prepared to live with, even augment, rather than give up what most of us enjoy. In the same sense it can perhaps be said that, for all our protests, we are prepared to live with crime.

For all this, we have still a long way to go before we come to terms with the reality and obduracy of crime. A major American Commission, promoted a few years ago by the Department of Justice to examine the "criminal justice goals", said many good and wise things. Yet its third report, issued at the beginning of 1973, stated flatly that the United States could and should reduce the rate of "high-fear" crime (murder, rape, aggravated assault, robbery and burglary when committed by strangers) by fifty per cent in the next ten years. The primary means to this were to be the prevention of juvenile delinquency, improvement of social services, speedier justice and greater participation by the public. Again, all excellent things in themselves, but scarcely any easier to achieve than a reduction in crime as such. No wonder that, at the end of 1974, there appeared, by way of addendum, the following verdict from the Director of the National Institute of Law Enforcement in the Department of Criminal Justice: In spite of soaring outlays no substantial reduction in crime was to be expected in the next five to ten years. That comes closer to sober realism.

The whole business of analysis is far more complex than was once imagined. Even when it seems obvious that certain conditions are conducive to crime, we have to recognise that social forces are far less malleable than was once hoped by radical and liberal reformers. Certainly societies are very different now from what they were a hundred or even fifty years ago. But the effects of the changes, often the changes themselves, have not been what their initiators designed or expected. Crime, in particular, has a way of feeding upon change, reappearing in new forms as old forms fade, battening on the characteristics of any society. As to what determines the shape of crime, Tarde's arresting aphorism still holds good: "The individual bestirs himself, society sweeps him along."

More plausible, at first sight, seems the suggestion that we should be able to take action to forestall trouble from individuals whose early characteristics warn us that they may prove specially prone to crime. Two American criminologists, Professor Sheldon Glueck and Dr. Eleanor Glueck, have devoted a lifetime of research to tracing and analysing criminal careers. They have sought to identify criteria, both genetic and environmental, which will enable people at particular risk of delinquency to be identified as small children and given special attention by welfare and educational services. But it is one thing to single out children as needing special assistance because they are at risk in terms of physical disease, quite another to single them out as likely to become persistent offenders. Even selectivity in terms of education has been condemned as a self-fulfilling prophecy. And when the Gluecks claimed, some years ago, that they had reached a

point where they could pick out, with reasonable certainty, the
potential delinquents amongst young children it aroused a storm
of dissent. Some people questioned the efficiency of their criteria.
At least as many people deplored the idea of applying them in
practice, on the ground that to do so would be socially divisive,
infringe human rights and produce more delinquency than it
could avert.

That drives us back to people in whom the risk has become a
reality, the illness or abnormality manifested itself in crime. Even
assuming that we could diagnose reliably, could we effectively
treat? Is it feasible to change characteristics that contribute to
crime? What devices are at our disposal, and how far are we
prepared to go in using them?

Techniques like psychotherapy and the use of drugs, which
can do much to relieve some kinds of mental illness, have so far
achieved little success in checking crime. When drugs do work it
is in a very narrow and specific way, as in removing the motive
for certain sex offences. The fact that in most cases the links
between mental illness and crime are diffuse and indirect makes it
impossible to guarantee, even where the illness can be controlled,
that the crime will disappear also. The habits associated with
criminality may have become too deeply embedded. And the
crime may not be connected with the illness at all. The mentally
abnormal may commit offences for the same reasons as the mentally normal and to treat their afflictions will not necessarily prevent them from continuing in crime.

Chemical manipulation of behaviour is still largely in the realm
of science fiction. Tranquillising drugs have proved useful in
quietening excited or deluded psychotics. But, except in doses
that make the subject sleepy or paralysed, they have been of little
help in reducing the ferocity of obstreperous prisoners. New
drugs are being produced at a great rate, and it is always possible
that something will be discovered that has much more direct and
dramatic effects upon mood and behaviour than anything in current use. If so, its effects would be unlikely to stop short at
inhibiting crime, any more than the effects of leucotomy stop
short at inhibiting depression.

Certain kinds of surgical intervention, once widely canvassed,
have been largely abandoned both as unethical and as ineffective.

Leucotomy, a form of brain surgery which it was once hoped would reduce tendencies to violent crime, has been found useless for that purpose. At the same time there has been revulsion against such a measure, except as a last resort to counter desperate mental suffering. Crude beliefs that castration could be employed to check violence or other kinds of crime have proved to be groundless, whilst drugs have taken its place as a safeguard against sexual offences. At the same time castration has come to appear as obnoxious as other old-style mutilations.

A more recent candidate as a miracle cure for criminals has been aversion therapy, the technique of getting a patient to associate some behaviour he wants to be rid of with, say, an unpleasant nausea. That has the drawback of being rather specific in its effects: you may be able to induce disgust for the stealing of chocolate, for example, but hardly for stealing generally. Moreover experience has shown that effects produced in this way soon wear off. And many psychiatrists have turned against the use of such methods because they dislike being called mere punishers. The opposite technique of rewarding certain kinds of good behaviour to the point where they become automatic has been found, in general, a more successful way of changing attitudes, and more lasting in its effects. It has been possible to use it, to a limited extent, in penal institutions. Then there is the much broader idea that the best way to keep people out of crime is to offer them something better. But, as Jeremy Bentham observed a long time ago, there would be strenuous opposition to a general system of paying offenders who kept out of further crime. Bearing in mind the disabilities and defects of many persistent offenders it is difficult even to set before them an alternative way of life within the law that will give them as much satisfaction, or make as few demands, as their current criminal ways. Without inducement to change why should they bother?

The social-psychological principles of offering them relationships with people they will come to trust and admire, whose guidance they will accept or whose example they will follow, can also founder on the rock of their personal peculiarities. The more detached they are from normal human relationships, the lower their capacity for trust, the greater the distance between their existing outlook and that of the therapist or social worker, the less the chance of achieving the vital relationship. Many of those who break the law, even repeatedly, are not like that, but it is precisely the hard core who are the least accessible.

Similar problems face us if, recognising that major change is unlikely or at best will be slow and uncertain, we try to find ways of propping up persistent petty offenders, protecting them from provocation or temptation. We know, for example, that socially inadequate offenders often keep clear of trouble as long as their parents are alive to protect them and may have long interludes free of crime later in life if taken under the wing of a sympathetic landlady. But their various unpleasant characteristics make it hard for them to find or retain such protection. Nor do hostels offer a sovereign remedy.

Institutions designed for the treatment of the mentally ill or retarded find criminals awkward customers, likely to upset other patients and staff, likely to abscond. There is a continual passing of the buck between the medical and penal systems, especially with regard to those labelled psychopaths. And when it comes to compulsory detention, the concept of psychopathy is not only a will-o-the-wisp but a dangerous one, less likely to lead to a pre-scription for treatment than a verdict of despair. Once an of-fender has been so classified further consideration of mental state or social circumstances may be curtailed. In some countries the law has allowed detention of psychopaths for periods far beyond what can be justified in terms of the possibility of treatment, the risk of danger to others or the limits of justice.

It is easy to ask for general explanations and expect straight-forward remedies. But the more we learn and experiment, the more we are driven to recognise the limits of our power to con-trol either society or individuals.

Part III

THE RESPONSE
OF THE LAW

CHAPTER

5

Defining Criminal Behaviour

The Constant Core of Criminal Law

Many have tried to find a single comprehensive definition of crime. No one has produced an answer that will stand up to scrutiny. You can say it is something that threatens serious harm to the community, or something generally believed to do so, or something committed with evil intent, or something forbidden in the interests of the most powerful sections of society. But there are crimes that elude each of these definitions and there are forms of behaviour under each of them that escape the label of crime. The argument that crime is anything forbidden, or punishable, under the criminal law is open to the objection that it is circular. But it is at least clear cut, it refers not to what ought to be but to what is, and it is an essential starting point whether we want to make comparisons or to consider whether criminal law today needs to be expanded or contracted.

Can the clue to what we consider a 'real' crime be found in the intensity, the unanimity of feeling about certain kinds of behaviour? Do we make crimes only of things that society as a whole views with strong indignation or apprehension? The Italian criminologist Baron Raffaele Garofalo concluded that the 'natural crimes' were those that caused deep offence to the sentiments of pity and probity prevalent in all classes of a particular society. Emile Durkheim likewise believed that universal disapproval was the basic element which made an act a crime, though he could not accept that the feelings which lay behind such disapproval were necessarily related either to the genuine interests of society or to justice.

Some people contend that the law, including the criminal law, is simply imposed from above by those in power, whether in their own narrow interests or because they feel better placed to judge that particular acts or omissions, apparently harmless in themselves, are in conflict with vital interests of the community as a whole. Certain jurisprudential writers, foremost amongst them the great nineteenth-century English legal philosopher Austin, have pressed the argument that crime is anything proscribed under threat by the sovereign, or most powerful in a particular society. Thus highly respected experts in jurisprudence see law as a series of general commands issued by the sovereign and backed by coercive threats. A crime, it follows, is a breach of such commands. One of the analogies used to make the point has a curiously radical ring: the coercive command of the sovereign is compared to the orders given to a bank clerk by a robber with a gun. What is defined and punished as crime, in other words, depends on who holds the gun, the power to assert sovereignty and the power to enforce it. That situation can be seen most vividly in terms of countries that have been conquered or occupied, that are ruled by dictatorial government, or where a minority have a stranglehold.

Many modern sociologists would go further, claiming that the key to what is classed as criminal is to be sought not in the common conscience and interests of a society but in the conscience, and still more the interests, of the ruling or influential classes. On this interpretation the criminal law appears as a weapon to maintain the domination and privileges of the few. An extreme example was old style slavery. Under such regimes the crime is not for the owner to assault or kill a slave or to deprive him of all he needs, but for the slave to resist or abscond. In medieval England the Norman conquerors made it a capital offence for a Saxon to be caught poaching or be found out of doors after curfew. In South Africa it is a crime for a black man to live in a white area. By analogy, goes the argument, modern laws of theft and damage and trespass are basically designed to protect the property of the rich from the depredations of the poor. There is an element of truth here but also an element of falsehood. The poor and the oppressed are more likely to find themselves victims of assault or murder, robbery or rape than the rich and powerful.

Closely linked with the concept of the law as the creature of powerful vested interests rather than the protector of the rights of all is the claim that groups and individuals are entitled to disobey laws they consider unfair or oppressive. Where this means that people need not keep laws which conflict with their personal or sectional interest, we are on very dangerous ground. We cannot each be judges in our own cause. Where it means that we need not, or should not, keep laws that outrage our deepest moral convictions or our beliefs about what is right for the community as a whole, it is a rather different question. Slavish obedience to evil laws has produced some of the darkest pages in history. And we have the famous instances in which behaviour classed as criminal in its own time or place has exemplified the height of moral heroism. Whereas a very great deal of wickedness never comes within the ambit of crime, there are a few crimes committed only by heroes or saints. You have Socrates condemned for corrupting the young, Antigone for carrying out her religious duty, martyrs slaughtered for refusing to deny their faith, Germans hanged for hiding Jews under the Nazis, Russians consigned to labour camps for demanding freedom of speech and expression. But these are the exceptions, the only way out where no other kind of protest or hope of change can be found. Moreover they all entail acceptance of the penalty for the crime. Indeed it is from this acceptance that they derive their strength, their disproportionate power to convince others, even eventually to transform the law.

It would be strange indeed if what was classed as crime were exactly the same in every society. But it would be stranger still if so potent, so elemental a concept were purely artificial, unrelated to the basic human struggles which it everywhere reflects.

Imagine that I settle in West Germany, a prosperous capitalist country, and there commit a number of crimes. After a few months, feeling things are getting too hot for me, I slip over the Berlin Wall and renew my nefarious activities under the red flag. I should, of course, have been warned that I would be entering an entirely different type of society and must forget my old bourgeois ways. But, if I went on committing the familiar offences, the thefts, frauds, robberies, assaults and murders, they would still be recognised as crimes both by the victims and the authorities. I could press on to the Soviet Union, to India, or the emerging states of Africa, to China or Japan, to Latin America, even to Alaska or Greenland, without escaping these inexorable threats to my freedom of action.

Obviously in some of these places I would face additional hazards. If I expressed my criticisms or my prejudices as freely in Russia or China as I do in England I could find myself charged with counter-revolutionary crimes. Indeed, in the unlikely event of my becoming a citizen of either of those countries, I should be committing a crime if I tried to leave it without permission. A capitalist state, in contrast, does not mind where I take my person, stipulating only that I do not export much of my property without getting leave. These differences flow from fundamentally different views of the relationship between the individual, the state and the ruling regime or ideology. Of all major crimes it is crimes against the state or the ideology that are most variable. Dissidence, even without violence, appears a major threat in a totalitarian country, a necessity to be tolerated, if not encouraged, in a democracy. Religious offences like witchcraft and heresy once took pride of place amongst major crimes: they have disappeared from modern criminal codes. About the more direct physical assaults on the persons and property of individuals, on the other hand, there is virtually world-wide unanimity.

Just as a modern traveller finds much in common wherever he goes, yet catches fascinating glimpses of different ways of life, different historical relics and tradition, different outlooks and attitudes, so it is with crimes. Some things I could safely do at home would be classed as criminal offences elsewhere; some distinctions as to gravity would disappear or be reversed; some excuses that would help me little in England might materially reduce my offence abroad, some, unimportant here, might aggravate it.

It is in the ways crimes are delimited, above all in the procedures for dealing with them and the penalties they carry, that the differences are to be found. The fact that I was arrested for my robbery or dangerous driving or drug trafficking, would not particularly surprise me. But the way I was interrogated, and the sentence I found myself serving if I survived, might well come as a very nasty shock.

Technically it may be true to say that any violation of the criminal law is a crime. But in common speech we mean by crime a more serious offence, something like what used to be called a felony in England and still carries that label in the United States. For lesser breaches many other terms have been devised in different countries: infractions, violations, délits, misdemeanours. Their minor standing is reflected in the procedures for dealing with them, like on-the-spot fines by police or adjudication by summary or peoples' courts, and in the lesser penalties, often no more than fines, attached to conviction.

You can travel from country to country, comb through one penal code after another, and find little substantial variations in the basic crimes. One reason is that we are becoming a single world in terms of crime as surely as in terms of economics or communications. Almost everyone has a criminal code marked by similar influences. For example, the Japanese readiness to study and adapt the experience of other nations in industry and commerce has extended also to law, including criminal law. They borrowed from China, then looked further afield to England, to France, and eventually, in 1907, adopted a penal code very similar to that of Germany. In one sense this switch from the approach of the far east to that of the west could be seen as a legal revolution. Yet a perceptive Japanese lawyer warns against over-estimating its impact, since "the basic concept of justice and the thoughts pertaining to right and wrong as they existed in pre-reformation days of the Meiji were not so different from those of the Western world." The China of Chairman Mao, often quoted as having no criminal code in the sense understood elsewhere, takes for granted the criminality of such offences as homicide, rape, robbery or arson, since "the seriousness of these acts themselves is very obvious."[15]

The Limits of Intervention

Political, social, economic transformations tilt the balance of criminal law as of everything else. The core of crime, like the core of human needs and greeds, remains the same, but the borders move. Is the criminal law now getting out of hand, spreading far beyond its proper sphere, and thereby ceasing to be either enforceable or respected? Or is it suffering from the opposite weakness, over-permissiveness, a withdrawal of established restraints, a betrayal of values that need its defence? Or again, can it be that nowadays we are getting both at once, like simultaneous inflation and stagnation in the economic sphere?

The traditional liberal position has an appealing simplicity. In the constantly quoted phrase of J. S. Mill, "the only purpose for which power can be rightfully exercised over any member of a civilised community is to prevent harm to others. His own good, either physical or moral, is not a sufficient warrant."[16] Since the criminal law embodies, at least in a free society and in time of peace, the power of the State over the individual in its most extreme form, it ought to follow that it should be kept to the absolute minimum necessary for public order and safety. Above all it should not interfere in private behaviour.

But there you have it. What kinds of behaviour are wholly private, wholly without repercussions upon other people or on society? We talk of "victimless crimes," offences such as possessing drugs for one's own use, prostitution (as distinct from causing annoyance by soliciting); illegal gambling; smoking or drinking alcohol in public when under age. But the vagrant or drug addict may infect others by his example. If he ruins his health, mental or physical, the personal and financial cost falls on others. The prostitute, too, may infect. The drunk or the gambler may ruin his family as well as himself.

So Mill is begging the question. What has to be considered is not simply whether others are harmed by some kind of behaviour, but whether the harm is direct enough and serious enough to warrant bringing to bear the criminal law. It needs to be direct, since the criminal law must be able to identify clearly the behaviour and the person it is punishing. It needs to be serious since it has to be weighed against personal freedom. The people who say that the criminal law has, as its primary task, the maintenance of the moral values of a society, are saying, amongst other things, that certain apparently 'private' behaviour has grave repercussions upon institutions or concepts like the family, or the sacredness of human life, which they consider vital to survival.

It is not simply a question of whether the value we are trying to uphold needs and deserves the support of the criminal law. It is also a question of whether the law can realistically be expected to provide such support. How far will the law be defied by those at whose behaviour it is aimed? How far will the rest of the society concur in attempts at coercion? Life, limb and state or personal property can, in general, be defended against direct threats, because scarcely anyone disputes that they should be protected. But as soon as you get beyond that, into moral beliefs, estimates of what is or is not socially harmful, you begin to lose the near unanimity which seems essential to enforce the criminal law.

There will be influential groups wholly opposed to certain aspects of the law. There will be trade unionists who see unrestricted picketing as the working man's basic right. There will be businessmen who claim freedom of enterprise to establish monopolies. There will be homosexuals who claim rights parallel with those of heterosexuals. There will be drug takers to whom addiction is a symbol of liberty. There will be women (and men too) to whom abortion on demand appears another basic human right. And there will be a growing body saying the same of euthanasia for those who want it. The claim that the law need not be obeyed by those who disagree with it, whether for profound moral reasons or because they see it as threatening their sectional interests, has been gathering strength. There are, however, marked differences between countries in the proportions of their citizens who think that the law must be obeyed, right or wrong. The Germans, for example, still rank high in their support for obedience.

Totalitarian countries have taken very seriously the educative, as well as the directly deterrent, functions of the criminal law. They have used it deliberately in most areas of life, with the avowed intention of changing attitudes and of stamping out dissidence. That was true of Tudor England, of Nazi Germany and Fascist Italy, as it is of many modern socialist and nationalist states. But in all countries and in all times there have been those who resisted. We may recall prohibition in the United States, wartime rationing, the wartime defiance of collaborators with occupying powers, clandestine religious teaching and practice where both have been prohibited for many years. Criminal laws intended to redirect or transform public opinion have left a long trail of frustration. The evidence is that, in the short run at least, their impact on established attitudes is negligible. If there are to be any educative effects they are likely to build up very slowly, over generations rather than years or even decades, and then only if the change in the law is in harmony with the trends of change in society.

Statistics, whether of the number of people who have taken drugs, or of the number who condemn strikes, or think homosexuals should go to prison, cannot, in themselves, tell us what we ought to do about the criminal law. We need to take into account all the evidence we can get about how far existing law is kept or broken, approved or condemned. But we are still left with such questions as does the frequency of lawbreaking mean we need still stronger measures to protect a value which, though threatened, is indispensable? Or does it mean that we should abandon the attempt to enforce a law which is failing so often to restrain?

The principle that the State should keep out of our private affairs and the principle that the criminal law should support our moral values both beg the question. Each offers its element of truth. The criminal law is indeed too drastic a weapon to be used except as a last resort and with the maximum of caution and precision. The criminal law is concerned with protecting some of the most fundamental and deeply felt values of a society. But these maxims cannot be popped, side by side, into some infallible computer, to produce lists of just what should or should not be criminal. Certainly it appears much easier for the criminal law to confirm values than either to create or to change them.

Offences have multiplied as societies have become more complicated and government controls have increased. It is comparatively easy to pass criminal laws, a very different matter to enforce them. There is the cost, in money, manpower and material resources. There is the cost in surveillance, loss of liberty and privacy. There is the cost in human suffering, not only to lawbreakers but to their friends and families. And there is the cost in respect for the law. The more things are counted criminal, the more commonplace becomes the idea of crime. The idea of offending ceases to be as shocking as it once was. This is aggravated by the introduction of offences of strict liability, mostly related to administrative controls, where a person can be found guilty without having any evil intent or acting with what could be called criminal negligence. Again, the multiplication of offences beyond what can possibly be dealt with by police, courts and the penal system, inevitably entails uneven enforcement. Procedures are over-stretched, distorted, exposed to ridicule or contempt, engendering a sense of injustice and disrespect for the law.

If we are not to go on expanding criminal law as society becomes more complex, what other means can we take to control misbehaviour? Marxist idealists used to prophesy that the criminal law would wither away, along with the state and the other institutions required to control the otherwise unbridled egotism of capitalist societies. Less high flown was Durkheim's thesis that criminal law expressed the powerful values binding together the members of simple, solid societies. He believed that advanced societies, where different groups and individuals had very diverse values and functions, must depend increasingly upon the regulations and compromises of the civil law rather than the black and white prohibitions of the criminal law. The object would be to oblige offenders to comply rather than to make them suffer by way of punishment.

Some of this we can certainly see. Trade unions on one side, monopolies on the other, are nowadays in a position to hold society to ransom. Their assaults upon private and public property far outrun anything threatened by ordinary crime. But the legal restraints upon them, and the machinery for enforcing them, are almost entirely civil rather than criminal. In the United States, for example, though for some offences the two procedures are available side by side, criminal prosecution is very seldom used.

We need constantly to remind ourselves that the criminal law is not the only way, often not the most effective way, of influencing opinion or behaviour. In industrial relations, in business and commercial affairs, in the prevention of pollution and protection of the environment, it has been found preferable to provide alternative facilities and rely on other controls and civil remedies. Factory inspectors, for example, are armed in England with powers to take action under the criminal law, but they seldom use them, considering it more effective to work by advice, persuasion, and warning. Now it has been proposed that they shall have additional powers to order work to stop where proper safeguards are lacking, surely a more effective device than prosecuting and fining an employer, perhaps to the serious detriment of his business and his workpeople's jobs. What is wanted, especially in this whole field of administrative controls, is not punishment but compliance.

Enforcement by way of administrative action rather than by the criminal law of course carries its own risks. A totalitarian regime of whatever complexion, with powers of direction covering the employment, housing, privileges, residence, of all its citizens, can threaten to inflict penalties of the utmost severity without resort to charge, trial or appeal. It can withdraw a permit to work in a particular profession, even to work at all. It can forbid a citizen to live in the city which is his family home and send him off to a job in a remote area for an indefinite time. Even under more liberal regimes abuses and injustices occur. But whereas in a totalitarian state top administrators are likely to be militant party members, whom government will encourage rather than check, in democracies they themselves are subject to controls. There is the weight of public opinion, the curiosity of the press, normally some possibility of appeal. Where administrative controls are used in a totalitarian state, it is for political reasons, as a part of the machinery for concentrating power in the hands of the party. Although bureaucratic systems everywhere prefer to keep the reins of power in their own hands, it is the essence of liberalism that they should be subject to checks. It will not help us to curb the excessive expansion of criminal law at the expense of an uncritical enlargement of the powers of administrators.

On this practical level we may well conclude that the criminal law should be held in reserve, as the very last resort, in securing compliance with necessary administrative controls, but that administrative decisions should themselves be subject to more of the checks, the publicity, the appeals that help to protect the individual under criminal procedure.

Hesitant Contractions, Sweeping Expansions

The acts we have primarily in mind when we think of contraction of the criminal law are such things as suicide and attempted suicide, homosexual behaviour between adults, adultery and prostitution. All these are forms of behaviour which have carried strong religious taboos, once widely influential, now ceasing to claim general allegiance. In the case of attempts at suicide, the impulse has changed from punishing to helping. In the sexual offences there has been franker recognition of the sheer impotence of the criminal law in face of behaviour that is persistent, widespread and deeply rooted; of the injustices involved in haphazard or discriminatory enforcement; of the dangers of blackmail; of the inevitable invasions of privacy in any endeavours to obtain evidence. On balance, the dubious effectiveness of the criminal law in restraining these kinds of behaviour has been outweighed by the indubitable social harm implicit in retaining them.

Abortion has been seen as in a similar category, subject to similar objections with, in addition, the physical danger to the woman who goes to a back-street abortionist. On the subject of euthanasia, the 'mercy killing', at his own request, of a person who is in great pain or slowly dying, there has been growing controversy. There are many who still see it as a sin in religious terms, or as at odds with the primary duty of the medical profession. There are others who take the line that, since we have now the power to prolong life, and with it pain and infirmity, we should accept responsibility for meeting requests to end it. It is arguments of expediency, however, that are coming to the fore on both sides. There are doubts about possible abuses by those looking after the sick, about intolerable pressures upon invalids who feel they are a burden, even about opening the doors to the kind of thing that happened in Nazi Germany. There are arguments on the other hand that euthanasia is common in practice where patients are dying in pain and that therefore it should be legally recognised and controlled. So far, whatever may be going on in secret, no country has dared to take the step of making euthanasia non-criminal, although a few put it in a lower category of homicide than other kinds of deliberate killing.

Even where the criminal law has been narrowed it does not mean that controls have been wholly withdrawn. Indeed in some fields, such as homosexual behaviour or abortion, it is claimed that relaxation of the law leaves other influences freer to operate, since people need no longer be deterred from seeking advice by fear of prosecution. Moreover some legislative controls are normally retained as a bar against exploitation. Though in England attempted suicide is no longer an offence, it is still a crime to incite someone to suicide or abet him in it. Though prostitution is not criminal, it is still a crime to traffic in women, or for a prostitute to pester people on the streets. Though homosexual behaviour between adults has ceased to be legally prohibited, it is still a crime to indulge in it in relation to juveniles. It is still a crime to carry out an abortion without complying with the criteria and regulations laid down by law. It has repeatedly been found, however, that such protective regulations in these spheres tend, in the first instance, to be loosely drawn. The sharks move in to exploit loopholes and weaknesses in enforcement, and further legislation becomes necessary to curb their lucrative depredations.

Relaxations and projected relaxations in criminal laws relating to pornography show rather similar trends. They reflect the loosening of religious restraints, greater sexual permissiveness, more concern with the right of adults to decide for themselves in such matters. It cannot be denied that a proliferation of gross obscenity lowers the whole tone of a community. But, in a sense, pornography is as basic as sex, existing in all societies. Moreover, its definition is very fluid, its boundaries virtually impossible to fix, and attitudes to it are very variable, not only between cultures but between groups and individuals. Movements by particular groups forcibly to control the sexual morality of adults, without reference to genuine harm done to others, have been worse than self-defeating, adding fuel to intolerance. The balance of experience and evidence so far does not support the idea that pornography fosters crime, even sexual crime. This is a sphere in which the intervention of criminal law should be reduced to a minimum, though full play should be left to other social influences such as homes, schools and churches. The law should restrict itself to the protection of the young and to preventing pornographic material being forced upon those who do not want it, whether by general advertisement or display or by indiscriminate use of post or telephone. Retention of these safeguards has been accepted as necessary even in avant-garde reports such as those of the Arts Council of Great Britain (1969) and the United States Commission on Obscenity and Pornography (1970). And they remain on the statute book even in countries like Denmark which have very permissive legislation on the subject.

Gambling is another offence on which the grip of the criminal law is slackening, not least because of the impossibility of general enforcement in face of the determination of substantial sections of the public to carry on with it, legally or illegally. In England we have moved from prohibition to licensing and regulation. In parts of the States, where illegal gambling has ranked as the most profitable of criminal rackets and the most widespread factor in police corruption, one form after another has been legalised, and state lotteries have moved in, adopting the philosophy that if gambling is inevitable the public purse may as well rake in the profits. This good-out-of-evil philosophy is accepted in several countries. In England it has so far been rejected, as implying outright encouragement of gambling, but the licensing of betting shops alone has conferred respectability upon it. The criminal law confines itself to breaches of licensing rules or to what amounts to fraud.

The subject of drugs has stirred stronger and wider controversy, largely because it has been affecting the young, and the young of all classes. Leave aside, for the moment, such long-established indulgencies as tobacco and alcohol. The history of drugs and the criminal law during the present century has been curious. At the beginning of the century the problem in North America was hard drugs: opium, morphine, heroin, cocaine, derived largely from medical sources and used mainly by adults. That was checked by stricter self-regulation in the medical and pharmaceutical professions and by stricter laws on labelling and dispensing. The criminal law was brought in to suppress the more public use of the drugs by groups of immigrants in the United States.

Marihuana, when it came into use there, was classified with the hard drugs and penalised accordingly. Now that situation has come under challenge for several reasons. The majority of those who have used the drug, at least in the forms and concentration generally available in the western world, do not appear to have suffered any serious lasting harm. Large numbers of young people, having discovered this and broken the law with impunity, have further lost respect for the law. Some have undoubtedly been thus encouraged to assume that legal prohibitions of more harmful drugs can safely be ignored as well. Some, on the other hand, have found themselves dealt with as criminal for taking marihuana alone. At the same time the idea that the criminal law can be used to oblige addicts to accept treatment and to achieve permanent cures has been shown to be misguided. Whether in prison or in hospital there is a dearth of skilled treatment and in any case it is impossible to get far with an unwilling patient or to prevent an immediate relapse on release. For all these reasons there are the insistent calls to 'legalise pot'.

Yet it is not quite as simple. Evidence is conflicting as to whether a minority, especially amongst those who use marihuana frequently and over a long period, do suffer serious effects. There is some reason to believe that it delays the process of growing up and coming to terms with reality amongst adolescents: yet it is adolescents who are becoming increasingly involved in its use. Whatever people may believe about the right to experiment with it, not many would advocate, as a Canadian Committee put it, that it should be freely on sale at every candy store. Nor can it be assumed that the only people who want to retain criminal sanctions against the use of cannabis are people remote from the problem. Leaders of black communities in the cities protest that to legalise pot would be interpreted as a design to keep the blacks enslaved. Certain developing countries, where drugs of this kind are a far more serious threat and the struggle against their misuse still more difficult, may feel a similar bitterness.

It is my impression that in Western Europe and North America pot has already passed the peak of its fashion. Though we cannot say that a definite downward trend has been established there are signs that the drug is losing its status as a symbol of youthful revolt. I would abolish criminal penalties for the simple possession and private use of cannabis in countries like England, Canada, the United States. That would help to split the drug culture, make it less likely that cannabis takers would be brought into contact with those on drugs like heroin or L.S.D. At the same time, however, a sustained system of education and warning should be provided. The Canadians have suggested that, whilst taking cannabis should cease to be criminal, public use should be met by confiscation to make it clear that the drug is still suspect and its use and passing on to be discouraged. Another suggestion is that it should be subject to licensing laws like those for alcohol,

controlling the outlets, quality, conditions of sale, perhaps prohibiting sale to people under eighteen. On the other hand, suggestions that manufacturing and selling should be criminal when
use is not, seem to invite trouble. That was tried with alcohol
during the prohibition era in the States and simply encouraged
criminal rackets.

There remains the problem of people addicted to heroin or
other dangerous drugs. For them I would consider registration
and strictly controlled prescription as preferable to the machinery of criminal justice. Unlicensed large-scale trafficking in
such drugs should be met with the utmost severity, as a serious
form of organised crime.

Another controversial section of criminal law is concerned
with disorderly conduct or suspicious circumstances, like wandering without visible means of support, loitering with intent to
commit a felony, begging, being a common prostitute, drunk and
disorderly, or sleeping out without permission. Such laws raise
the old and basic conflict between the need for police powers and
the protection of individual liberties. They are less definitions of
crimes in the ordinary sense than powers conferred upon police
to take action against people who may be seen as a threat to
order, at most as potential criminals. They give very wide discretion to the police, allowing them, in effect, to by-pass the
principles of criminal justice. Vagrancy laws have enabled the
police to arrest, search and hold people they suspect of other
offences, where they would otherwise have no authority to do so.
"Disorderly conduct" has given them a handle for action against
people who insult the police, or against pickets or other demonstrators, who, it is feared, may overstep the bounds of legality
but who have not yet done so. The police claim these laws are
essential to enable them to deal effectively with potential criminals, to forestall crime or disorder. The suspects see them as
unjust, allowing arbitrary and oppressive action, encouraging
double-dealing by the authorities, in some cases making a crime
of poverty and its consequences, increasing hostility to the criminal law and those who enforce it. I would agree with those who
want to abolish them all, substituting more specific police powers
to enquire, warn, and, if necessary, escort out of the area individuals or groups involved in threatening behaviour. Something similar, combined with drying-out and treatment centres, may prove

the best answer we can find to the obdurate problems of the drunk and disorderly. The criminal law should be reserved for offences of persistent disobedience, active obstruction, or the kind of violence or threats against the persons or property of others that would be offences in their own right.

Suggestions that certain kinds of theft might be taken out of the criminal law may sound revolutionary. Some European countries, amongst them Norway, Sweden and Switzerland, already come very near to it. In Norway, for example, there is explicit exemption from prosecution if petty pilfering is from a relative with whom the offender is living. Colombia pushes exemption still further and at the same time spells out the conditions in more detail. To be held exempt a thief must have been impelled by a pressing need for food or clothing for himself and his family, he must have had no lawful means of satisfying it, he must take no more than is necessary, he must not be dangerous to society and he must use no violence. In other words, he must conform to everybody's picture of the poor man who steals to feed his family when every other alternative fails him. Many simpler communities, from the Pole to the Equator, take a lenient view of petty or isolated thefts, but judge the persistent thief a social danger who must be severely punished, even eliminated by death or banishment.

The idea has recently been mooted that, in face of the vast number of thefts already being committed with impunity, and of the pressures on the system of criminal justice, the criminal law ought to be realistic, accept the fact that many thieves are not and cannot be prosecuted, and limit its sanctions to the more serious and persistent. But there are substantial countervailing considerations. Theft is certainly not a "crime without victims." It would be very difficult to draw the line in terms of harm done: even a very minor theft can cause great distress. The loss of a radio or of some small sum set aside to meet bills for rent or for heating can constitute a major disaster for the old and poor. A recent survey in the United States has brought to light that over half of the oldest people questioned had felt obliged to limit or change their ways of life to reduce the risk of becoming victims of crime. Moreover a minor infraction left unpunished can act as an incentive to more serious stealing. And a general exemption of the kind proposed could not fail to affect general deterrence. I would say that, in this instance, we cannot contract the law: we must rather depend upon wise discretion in its operation.

The contraction of criminal law, particularly with regard to behaviour which has come to be regarded as a matter for individual moral choice rather than state prohibition, has been very marked in England, especially in the last two decades. The same has been generally true of the Nordic and one or two other Protestant European countries. Even so there are exceptions. Homosexual behaviour between adults, for example, is still an offence in Norway. And the bulk of western Europe has remained resistant to change. In France, for example, a wife can still be sent to prison for adultery, a husband for keeping a concubine; and in spite of widespread and publicised defiance of the law on abortion it has proved a most painful and tortuous task to secure its relaxation. In the United States and other parts of the English speaking world there have been strong pressures for the reduction or elimination of so-called 'morals' legislation; there is flagrant non-enforcement. But there has seldom been repeal. In many parts of the States, for example, adultery and simple prostitution are still offences under the criminal law. In less developed countries, in South America and parts of Africa and Asia, criminal codes still reflect religious taboos. Where a new upsurge of nationalism is taking over they may even be reinforced. In Kenya not only adultery but enticement of another man's wife ranks as an offence. Elsewhere the threat of imprisonment for wearing a mini-skirt in public may be a bizarre way of asserting independence and moral superiority.

Socialist countries have their own brands of militant puritanism. And unlike more permissive regimes they have no compunction about limiting individual liberty. On the contrary, the suppression of non-conformity, whether in sexual matters, in work, in the dissemination of ideas, fits well into their regimes. Most, for example, class homosexual behaviour as criminal even whilst they reject the religious tenets often held to underlie its proscription. They outlaw not only begging but incorrigible unwillingness to work, not to mention attempts to make money by speculation. And they use their own brand of vagrancy laws not only against down-and-outs but against 'parasitic' artists and writers who refuse to toe the party line. Switches of policy on matters that would, in the West, be considered fundamental issues of human rights, are made on purely utilitarian grounds. In Hungary, for example, within a short period of time, abortion has been made legal and then illegal again in response to fluctuations in population.

With a few distinguished exceptions, those who write about the contraction of criminal law are those who favour it. But the impression they create is misleading. If we differentiate, as we should, between the ideas of advanced jurisprudential writers and the law as it is, something very different emerges. The trend towards contraction has been confined to a very few countries which combine strong liberal traditions with a measure of basic stability. Most parts of the world have neither. Under the pressures of vast problems of change and disorder they have been moving in the opposite direction.

Even in the most liberal countries there have been inexorable pressures towards expansion.

Humanitarian concern has produced laws penalising neglect, abuse or exploitation of employees, cruelty to children or to animals. Welfare provisions have brought corresponding offences of fraud or of failure to contribute.

The complexity of modern social administration, the intervention of the state in many spheres of life, has added an enormous mass of offences to the quota. Evasion of duties imposed upon us, of taxes or licence fees, can bring almost anyone within the scope of the criminal law. Well over half the cases before our criminal courts deal with motoring offences: a whole new category of crime stemming from the scale and speed of transport and the attempt to prevent chaos on the roads. Of course you could always be charged with murder in the old days if you ran a man down with your horse or your carriage or your farm cart but the threat to life, property and order from such sources was negligible compared with what we face now. In sheer numbers the administrative and traffic offences created within the last century completely outbalance the varieties of traditional crimes.

The unprecedented expansion in science and technology has made the traditional protections offered by the criminal law seem almost impotent in the face of far reaching new threats to personal privacy, reputation, property, life and limb. We have always been vulnerable to defamation, and traditionally defended by laws, both criminal and civil, penalising slander and libel. But

what of pervasive new threats to reputation, much harder to
identify or trace to their sources, in centralised records or com-
puters? Our privacy has always been protected by laws against
trespass into our homes. But what of devices that make it possible
to record conversations from a distance? Our pockets have been
protected by the law of theft and the reasonable precaution of
keeping most of one's money in a bank. But what of the new
possibilities of fraud opened up by credit cards? Driving without
due care, an enormous category of offences, has been brought
within the scope of the criminal courts. But what of putting on
the market, without due care, drugs that may cripple unborn
babies? May organ transplants raise new issues related to manslaughter? And how can the old fashioned rules of war, forbid-
ding slaughter of unarmed civilians, stand up to the cold, remote
destruction of whole cities by the latest in bombs, of whole populations by germ or chemical warfare? These modern scientific
crimes have a scientific quality of detachment, of generalisation,
almost of inevitability, that make attempts to outlaw them seem
futile.

The ease and speed of travel and communication has also contributed to legal restrictions. It is a crime to smuggle immigrants
into Britain from the Commonwealth. It is a crime to smuggle
emigrants out of Russia to Israel. It is a crime to broadcast from
an unlicensed radio station or possess an unlicensed television set.
The hi-jacking of aircraft has added a new, intransigent, dimension to piracy. The betrayal of industrial secrets to foreigners is a
special category of crime not only in the iron curtain countries
but in France.

Over much of the world powerful secular ideologies, the varieties of communism and the new manifestations of nationalism,
have taken over the role of the traditional religions. They too
produce their idées fixes. There are whole new categories of
offences classed as counter-revolutionary or as crimes against the
socialist economy: the crime of the manager who, by neglect,
causes serious waste of money, material, energy or labour; or
even just displays "an unconscientious attitude towards the protection of state property". The criminal law expands rather than
contracts, in the endeavour to enforce the true faith, to correct
or make examples of those who betray it by lapsing into bourgeois heresies or failing in their socialist duties.

The growing assertiveness of minorities of all shades has contributed its share to the proliferation of criminal law. New ways to draw attention to grievances, new ways to plague the authorities, have gone round the world, from Gandhi's civil disobedience and its modern equivalents in South Africa to the English 'sit in'. We find in European criminal codes, on both sides of the iron curtain, such widely phrased offences as 'taking the law into your own hands' or 'arbitrary conduct'. Even liberals, traditionally wedded to the principle of keeping criminal law to a minimum, have added their quota, successfully advocating legislation to prohibit various forms of racial or sexual discrimination.

The Hard Road Towards International Criminal Law

The crimes of individuals, proscribed and punished under the criminal law of various nations, are a drop in the bucket when compared with the massive evils inflicted time and again by rulers upon those they rule, by one people against another in war or in subjugation. Most of these evils are outside the province of traditional criminal law, concerned as it is with regulating the behaviour of citizens of a single country, owing allegiance to a single government. They do not fit in with the conception of crime as an offence against the sovereign or the state, backed by a wide and well established measure of consensus, capable of strict definition and supported by sanctions which can be imposed upon individuals found guilty by the courts of their own countries.

Some progress has been achieved in international consensus. There are extradition agreements, albeit still very limited, concerning the return of people accused or convicted of certain basic crimes. There are a few offences, like piracy and slavery, of grave concern to humanity at large, which have been outlawed by international agreement and eliminated or drastically reduced by international cooperation. But by far the most difficult matters to agree upon, to define, to control, or to bring home to those responsible, are the actions taken by governments and nations, under the umbrellas of internal policy, of self-defence or of war.

These are not necessarily at odds with the internal laws of nations: they may be embodied in them. Guilt is diffused, actual perpetrators can often plead superior orders. What experience there has been of calling to account a nation's leaders in terms of war crimes has rested on the shaky foundation of retrospective legislation.

Piracy at sea, which used sometimes to be surreptitiously supported by governments as a handy weapon in their cold wars, has over the centuries been internationally outlawed and virtually eliminated. Slavery and the slave trade, once accepted as essential to economic and commercial life, have been similarly condemned, though evasions continue in practice. It is only slowly and with difficulty that nations are struggling towards agreement on defining offences of polluting the air, the sea, the rivers and the land. Suppression of the manufacture, sale and use of dangerous drugs, remains a major problem. Too many governments and vested interests—as well as too many crooks—have too much to gain. And the smaller though more dramatic problem of hijacking aircraft has demonstrated how difficult it is to reach international agreement on action even against a most dangerous offence which may strike at any moment against the interests of any nation as well as against innocent lives.

Two world wars in the present century, with mass slaughter of civilians, some on the orders of their own governments, have fostered the conception of 'crimes against humanity', actions which, even if legalised by a particular state, are so abhorrent as to constitute crimes in the eyes of mankind as a whole. This is a kind of secular version of the old claim of the church, as guardian of the conscience of Christendom, to condemn and punish offences committed even by rulers.

Rousseau was one of the first to challenge the assumed right to kill an enemy after he had surrendered or could no longer fight, arguing that he was then no longer an enemy, who must be exterminated in self-defence: he was an ordinary man again, and you no longer had any title to take his life. The voluntary initiative which established the Red Cross to care for the sick and wounded of all belligerents received immediate support from twelve states in the Geneva Convention of 1864, all of them promising to respect the immunity of military hospitals and care for the sick and wounded. In 1907 the Hague Convention went on to affirm the rights of all war prisoners, wounded or not. And after the first world war support for the Red Cross was written into the Covenant of the League of Nations.

It might seem that the slaughter and pillage of civilian populations should have been outlawed even more decisively than the killing or ill treatment of prisoners who have been combatants. In fact it has presented more obdurate problems. If you are trying to coerce a nation must you not attack the sources of its prosperity and supplies as much as its fighting men? The pressure to do so has increased rather than decreased in modern times. It is often said that modern wars are fought by civilians as much as soldiers: the equipment is at least as vital as the men and it is civilians who make the armaments. Moreover modern weapons of mass destruction do not discriminate, even in favour of babies. If all that was needed to make behaviour a crime were moral revulsion or social harm, every nation would have outlawed annihilation of cities from the air on the same principle as many have outlawed the use of gas or germs in warfare.

Some of the most horrible onslaughts of authoritarian regimes have been directed not against foreigners but against sections of their own people, supported by their own laws. It would take five hundred years of ordinary criminal statistics to match what has been done legally at their behest in a few months. But could a government in modern times be legally called to account for committing crimes not defined as such by its criminal code, crimes within its own jurisdiction? Could it be punished for such offences, and on what basis? The concept of 'crimes against humanity' and the verdicts of the Nürnberg Court were the answers.

Hitler's measures to exterminate whole ethnic groups in Germany and elsewhere left genocide the immediate and burning issue. In 1946 the General Assembly of the United Nations resolved that it was "a crime under international law, contrary to the spirit and aims of the United Nations and condemned by the civilised world." In 1948 a convention proclaimed that "at all periods of history genocide has inflicted great losses on humanity," and that "in order to liberate mankind from such an odious scourge international co-operation is required." Genocide, whether committed in time of peace or in time of war, was a crime which the signatories undertook to prevent and punish. It was defined as including attempts to destroy a national, racial, ethnic or religious group, whether by killing them, seriously injuring them, subjecting them to conditions of life likely to destroy them, taking measures to prevent births amongst them or forcibly removing their children. Those guilty were to be punished, whether they were private individuals, public officials or constitutionally responsible rulers.

States accepting the convention were to make genocide a crime within their own countries and to provide effective penalties. Those charged with the offence were to be tried either within the state where it was committed or by an international tribunal, provided that its jurisdiction had been accepted by the country in question. Any country entering into the convention could call on the United Nations to act for the prevention or repression of genocide. Disputes about the interpretation or application of the convention could be submitted to the International Court of Justice at the request of any of the parties to the dispute. Neither the exclusion of political offenders from extradition nor, under a later covenant, the time limits normally set for prosecution, should serve to enable those charged with genocide to escape arrest and trial.

There could hardly be a crime more horrifying, more patently harmful. Yet no more than half the states members of the United Nations have so far ratified the convention on genocide. What government likely to contemplate or wink at such an offence is going to submit itself or its officials to the prospect of formal conviction, let alone punishment?

Genocide, however, is by no means the only 'crime against humanity' proscribed by the United Nations. In 1945 its Charter affirmed co-operation in respect of human rights and fundamental freedoms. In 1948 its Universal Declaration of Human Rights declared that disregard or contempt for such rights had "resulted in barbarous acts which have outraged the conscience of mankind," asserted that they must be protected by the rule of law and that a common understanding of those rights and freedoms was essential. The civil and political rights enumerated were in the tradition of those long since affirmed by such national declarations as the English Bill of Rights and the Declarations of

Human Rights following the French and American Revolutions: basic guarantees of the liberty of the individual even against arbitrary oppression by his own government.

But again it is one thing to assert that it is an offence against humanity to pass retrospective legislation, arbitrarily to deprive anyone of life, liberty or property, to suppress freedom of speech or movement, quite another thing to get governments to accept the authority of an international tribunal to conciliate or judge in cases where they themselves stand accused. A few have submitted to this, in the knowledge that if other countries can lay complaints against them they can likewise complain against others. Even fewer have gone on to accept the right of individuals within their jurisdiction to seek justice, as a last resort, from an international committee or court.

When it came to setting up the machinery to implement these covenants there was much debate and disagreement. Principles are one thing; accepting reductions of sovereignty, binding countries to genuine international action is another. When the matter was under discussion Australia wanted an International Court on Human Rights, France an International Investigating Committee plus an Attorney General, India wanted the Security Council to be given jurisdiction both to investigate and enforce redress of alleged violations. The United States and the United Kingdom proposed, more cautiously, a Human Rights Committee to deal only with inter-state disputes. The Soviets were against any international arrangements which might undermine national sovereignty. Meanwhile, both the Europeans and the Americans have been drawing up their own lists of human rights, setting up their own machinery, very similar to that of the United Nations, to deal with complaints on a regional basis.

One major obstacle to the acceptance of international crimes, punishable under international law, is fundamental disagreement, in fact if not in theory, about the limits of the rights to be protected. Some of these divergencies became apparent as soon as the attempt was made to translate the broad declarations into concrete stipulations for the purposes of the covenants. For example, the imposition of compulsory labour is condemned, but subject to such exceptions as "emergency service" or "normal civil obligations." Restrictions on freedom of movement are condemned, but not when they are provided by law, or necessary to protect national security, public order, public health or morals.

Nations have a deeply inbred conviction that they are the best judges of their own interests and of the limits that should be set to their behaviour. Yet experience has been forcing upon them an uneasy, but growing, awareness that without some kind of international consensus and regulation their very existence may be at risk. However, in the absence of an effective and overriding authority, whether based on power, on consensus or on loyalty, can 'international' crime be defined and punished as such? Can nations be obliged to accept international judgment, let alone condemnation? Those most likely to accede to it are, unfortunately, those least likely to need it.

International criminal law, like any other criminal law, depends in the last resort on the willingness to accept and enforce it. Its development has been hampered by the evasiveness of interested powers, the diplomatic wrangles about conventions, the vast crimes against humanity committed under the veils of national integrity, independence, dignity. Yet we can trace a path, painful, slow and tortuous, towards the idea that some crimes affect all mankind and that all nations should unite to outlaw them. Whether we shall ever reach this goal is by no means certain.

CHAPTER
6

The Instrument
of Deterrence

The Nature of Deterrence

Suppose, suggested the wicked Voltaire, you could sit by the fire in Paris, a glass of brandy at your elbow, and just by pressing a button you could end the life of your rich old aunt in Cairo who had left you all her money—how many of you would not press the button? Some, indeed, would not. They are angels. Some would not only press the button but rush off to Cairo and strangle her as well, just to make sure. The rest would weigh the chances: some would do it only if they were sure they would not be found out, some only if they were sure they would not be punished, some if they were convinced the punishment would not be too severe.

Do most people calculate the risks like that? Can crime be prevented or reduced by more vigorous penalties more vigorously enforced? Are the angels so by nature or do even they owe something to the existence of penal laws? Worst of all, what is to be done about those at the other extreme, too mad, too resistant, too tempted or too confident to be deterred by the dangers either of detection or of punishment?

Deterrence is a confusing word. It may mean preventing an offender from repeating his crime, what is known as individual deterrence. It may mean preventing the rest of us from following his example, what is known as general deterrence.

Dr. Johnson defined 'to deter' as "to discourage by terror, to fright from anything". But nowadays we are more squeamish about the idea of suppressing misconduct, even crime, by naked fear. But can it not simply be assumed that, if the penalties prescribed for those who break the law are severe enough, they will be deterred by their punishment and the rest will be deterred by example, whether the example works by way of fear or by building up the conviction that behaviour so punished must be morally abhorrent?

Eighteenth century criminal law in England carried the theory of deterrence by severity to its logical conclusion. Society was changing rapidly, then as now. On all sides people were being uprooted, on all sides there were new opportunities and temptations to theft and disorder. Persons and property had to be protected. Masses of new offences were introduced into the criminal code, scores of them carrying the death penalty. In practice only a small proportion of those who transgressed them were hanged. The very disproportion of the penalty to the crimes proved self-defeating, for few would risk bringing a fellow creature to the gallows for a minor offence. Juries were reluctant to convict, judges to sentence, administrators to execute. Penal theorists tried to make a virtue even of that. It was cited as a prime instance of penal economy that "few actually suffer death, whilst the dread and danger of it hang over the crimes of many". But there was neither justice nor humanity in such a system. It was not even efficient.

Critics of despotic power and arbitrary punishment in Europe still accepted deterrence as the justification and object of penal policy. But in the name of efficiency as well as justice they made proportion, moderation and certainty their watchwords. If people were to be deterred from crime, they must know in advance what punishment they might expect for any given offence. Such punishment should be sufficiently severe to deter from the crime, but no more so. As such, it would not be repugnant to public opinion and could be imposed with speed and certainty. It would serve to check the would-be criminal, who could weigh his crime against its consequence. It would give everyone a clear understanding of what was forbidden and of the seriousness attached to each kind of transgression. It would accord with justice and the rule of law. Principles such as these distinguish between deterrence and terror.

In the first half of the nineteenth century, on the Continent, in the United States and in England, much labour was devoted to attempts to embody them in logical scales of crimes and punishments. Yet they too proved disappointing deterrents. Moreover there was growing recognition that exceptions had to be made. It was not, after all, so just, or so efficient, to tie the punishment rigidly to the offence. Children, young people and the feeble minded were obvious exceptions. And should first offenders of previously good character really be treated in exactly the same way as confirmed recidivists?

It is often argued that to take into account the personality and needs of the offender as well as the facts of the offence must inevitably weaken general deterrence and foster rising crime. Yet the kinds of offence that have been rising fastest, such as serious assaults, burglaries, robberies, frauds, are the kinds of offence for which severity of sentence is likely to be tied most closely to the gravity of the crime. It is in dealing with the vast middle range of lesser crimes that courts are most likely to make concessions to the characteristics and needs of the offender. Nor should it be forgotten that, for the criminal assessed as dangerous, this personal factor can lead to more prolonged detention than would have been justified on the basis of the crime alone.

General deterrence cannot be written off as a survival from less enlightened times. On the contrary it has of late been thrusting its way again into the forefront. We all owe a debt to Johannes Andenaes, a Norwegian Professor of Criminal Law, for his elucidation of this concept and for bringing it down to earth. The fact that we are recognising the difficulties of reforming those who have once got into the habit of breaking the criminal law, together with the runaway rise in crime, has been forcing us once again to ask how people are to be prevented from embarking upon crime in the first instance. We can neither hope nor wish to rely wholly upon police vigilance and instant detection. What, if anything, can be done to build up and support other, less tangible, defences?

General deterrence is as difficult to understand and achieve as the deterrence of the individual criminal. It needs to be examined just as critically, yet we cannot afford to ignore or abandon it. We may utterly reject the concept of a society ruled by terror. That does not mean we must go to the other extreme and deny the usefulness of an element of deterrence, even of fear, in the balance of human relations. Experimental evidence has been confirming what commonsense has always assumed. Whether with horror or with envy, we imagine ourselves in other people's shoes and are influenced by what we see happening to them. If they pay the price of their crimes, our resistance is strengthened, if they get away with them and enjoy the advantage over us, our resistance is weakened. Deterrence is an essential protective device.

The criminal law is not, of course, the only source of deterrence. Religion, social convention, the pressure of family or group, can all have a powerful influence in impelling people towards some kinds of attitude or behaviour, away from others. Nor do criminal sanctions operate in a single or simple way. The criminal prohibition—the law and the punishment attached to disobedience—is only the first stage. Each step in its enforcement carries its own potentialities for deterrence, or for failure to deter.

For some people it is enough to know that "there is a law against it": the law is to be obeyed and that is that. For most obedience is likeliest when they are in general sympathy with the law and its objectives and accept its necessity. Even so, many are not deterred from occasional breaches, which they see as venial, doing no one any harm. They would not commit murder, robbery, rape, or even steal from anyone they know. But they have little scruple about breaking traffic regulations, admittedly as necessary for their own protection as for that of others.

Publicity is obviously an essential element in deterrence, especially where the newer laws are concerned. Whether the mechanism is to be education or fear of consequences or both, there must be a knowledge of the law and the sanction. A survey carried out in 1967 in California revealed that the average citizen knew the maximum penalty for fewer than three out of eleven offences put to him. Judging from the recurrent public demands for severer penalties for violent crime it is very doubtful indeed whether the average Englishman is any more aware than the average American of the actual severity of the existing law. A perceptive study made some years ago by Professor Nigel Walker seemed to confirm such doubts. In fact there are few, if any, offences involving deliberate and serious physical injury to other people which do not already carry a possible sentence of imprisonment for life. In California it was found that prisoners were much better informed than most of the public about the penalties attached to various crimes. To at least half of them, however, this knowledge was important not as a deterrent but as a bargaining counter should they be so unfortunate as to be caught again.

The chance of detection—or rather the chance as seen by a person tempted to crime—stands out as the most crucial factor in the whole chain of deterrence. Where conscience alone is insufficient, the fear of exposure, rather than any consideration of the severity of possible punishment, is likely to prove the strongest barrier against transgression, at least for the normally law-abiding. And the amateur offender, unlike the professional, is likely to exaggerate also the probability of being caught. Perhaps it is a matter of learning, and getting used to, the occupational risks.

Speed, as well as certainty, enters into the deterrent equation. A threat or penalty which is remote in time (like that of lung-cancer from excessive smoking) has far less impact than one seen as imminent. If that is the reaction of the individual concerned it is also the impression made upon the public at large. The fact that a notorious crime goes undetected for a long while, the rumour that a successful robber has had several years living abroad, in luxury and sunshine, on his ill-gotten capital, makes an impression which even his eventual capture, trial and punishment cannot erase. The penalty, if and when it comes, has lost much of its connection with the crime and therefore much of its deterrent impact.

A demand for severer punishments is the most popular response to rising crime, atrocious offences or even hooliganism. If only we brought back the death penalty, or flogging, or simply make prisons less comfortable, it is assumed that we should protect old ladies from thugs and railway carriages from wanton destruction. Harsher penalties for the guilty would buy safety for the innocent.

But even severity of punishment is not a simple concept. There is the maximum penalty laid down by law. There are the penalties imposed by the courts, which, quite properly, reserve the maximum for the very worst cases coming before them. And there are the penalties actually carried out: death may be commuted to imprisonment, imprisonment shortened by remission or parole. An unrealistic severity in legislation is almost inevitably modified in sentencing, excessive rigidity in sentencing by extending administrative discretion. Quite apart from the question of whether severe punishments are more effective deterrents than milder ones, a very high level of severity cannot be consistently maintained, at least in anything like a free society.

One final element must be added to the deterrent balance. The stigma attached to arrest, trial, conviction, may have little effect on the enthusiastic juvenile delinquent, the habitual criminal or the dedicated protestor. But it is probably a more powerful deterrent than formal punishment to the average respectable citizen. The offender caught in crime for the first time is more likely to ask "Will it be in the papers?" than "Will they send me to prison?"

Responsiveness to Deterrence

Generalisations about deterrence are as full of holes as most other generalisations. Different people and different groups obviously respond very differently.

Voltaire's angels, devils and waverers all faced the same threat of execution for murder. Some did not even need to think about it, others ignored it, the rest weighed it up. A modern Frenchman has subdivided the central mass of waverers into the morally cold or lukewarm, who need the direct brake of punishment to keep them from overstepping the legal limits, and the large army of the moderately moral, who nevertheless need the support of the law to keep them so. For them legal deterrence is comparatively remote, but not for that reason superfluous.

What puts people into these various classes? Comparison of criminal statistics and a look at individual peculiarities has already given us some clues. Youngsters figure more prominently than their elders, men than women. One way of interpreting this is in terms of response to deterrence. Younger people are, in general, readier to question or defy authority, less likely to be held back by the consequences or to appreciate what they may be, less weighed down by responsibilities, more daring, less controlled. They are not only more likely to be involved in crime but they are more than twice as likely as older people to be injured in driving accidents. There is some evidence that men are more willing to take risks than women and feel less threatened by social disapproval. Addiction to drink or drugs can put people under pressures against which ordinary measures of deterrence are impotent. Some kinds of mental disorder or abnormality are recognised as impairing understanding or control. Certain offences, by definition, tempt only sexual deviants. There are even a few people who seek out punishment: what is intended to deter becomes for them an incentive. These states may be comparatively rare. But they are only the extremes of a spectrum which runs right through from the most reckless to the most timid, the most recalcitrant to the most conformist.

However, the attitudes of individuals, indeed, their very personalities, owe much to the groups to which they belong. And such groups carry their own deterrent sanctions, which may be more powerful than those of the criminal law. The approval or disapproval of those with whom you live, work, share your leisure, interests and affection, have a stronger impact than the remoter sanctions of the state, however impressive. The small group retains the homogeneity and immediacy which the large, remote modern society has lost.

Some such groups, like the family, the church, the school, have been regarded traditionally as disseminators and supporters of law-abiding behaviour. The habits and inhibitions inculcated in childhood have been relied upon as a massive first line of defence against crime. This indirect influence has been classed as one of the most important functions of the criminal law. Even these groups, however, have their own loyalties and priorities. Where a state turns the law against parents, churches, educators, these groups may reverse their supportive role and prove the most stubborn and unshakeable advocates of defiance. Under totalitarian systems it has often been they who have stood out against the most extreme legal sanctions. But in many modern states, whether industrially advanced or disturbed by the process of industrialisation, it is the weakening rather than the strength of the traditions of these groups that gives rise to anxiety. The less certain their voice, the less accepted their authority, the less they can serve as alternatives or supports, and the greater the burden of deterrence left to be carried by the criminal law.

There are other groups which tolerate, or take for granted, a certain amount of law breaking. It may be speeding or tax evasion amongst the better off. There are families, streets, even whole neighbourhoods, where juvenile delinquency, petty theft or burglary are commonplace. If such attitudes do not actively stimulate crime, they at least weaken the impact of legal deterrence. The offender who is caught is as likely to get sympathy as blame; even a prison sentence is deplored as a misfortune rather than a disgrace, and neither conviction nor punishment carries much stigma.

But there are also groups which actively encourage law breaking. Some are political or ideological. They glory in invoking arrest, trial, imprisonment: for them, too, deterrents act as incentives. They oppose their own counter-deterrents to compliance: the comrade who will not defy the law or its minions is a prig, a coward, even a traitor. Then there are the gangs of juvenile delinquents, so lovingly analysed by American sociologists. It may be going too far to claim that they set up standards in opposition to those of society as a whole, like juvenile Fausts crying "Evil be then my good". But undeniably they sometimes find courage in numbers, egg each other on to violence, mischief, assault or destruction, with the disregard of consequences that can sweep any excited crowd.

Much has been made of the idea that adult criminals too have their own codes and sanctions. Mayhew described them, in Victorian England, as a hierarchy of professional groups, each with its own skills and prestige, the successful safebreakers and robbers at the top. Modern sociologists have painted similar pictures of the society within a prison, the daring criminal genius standing out amongst his fellows by his very defiance of the law and its sanctions, his refusal to knuckle under. It has been suggested that some men in British prisons have refused to be considered for parole simply because they do not want to lose face before their mates. At the other extreme the greatest contempt and the strongest physical sanctions of the criminal underworld, are reserved for the man who co-operates with the police in detecting others rather than face punishment himself.

Not only are the offenders we most wish to deter the least amenable to criminal sanctions, but the same can be said of the offences. The problems are by no means the same for every type of crime. Preventing murder or rape is a different kettle of fish from preventing parking offences. The motive for the crime, the strength of the temptation, the strength of inhibitions or moral revulsion against it, all affect the response, quite apart from the risks of legal punishment.

Crimes involving personal violence often spring from a sudden impulse, perhaps of anger, perhaps of despair. Deterrence, in its narrower sense of fear of punishment, may not come into consciousness in time to hold back the blow, or the emotion may be so strong as to banish fear of consequences. But that is not to say that deterrence, in its wider sense, is not operative. Though we talk of modern society as violent, the unconscious restraints against physical attack are still extremely strong. Such crimes are a tiny minority of all those committed. The fact that the severest penalties of the criminal law have long been imposed for murder and crime involving serious personal injury has doubtless contributed to this restraint. A radical change in the rules, such as takes place in war, soon reveals the potential violence which is just under the surface. The revulsion against killing or mutilating other people is overcome by most men when they are told the rules are different.

It has often been suggested that the offences least susceptible to deterrence are those that give direct expression to some strong personal need or emotion. That would include crimes of violence, sexual offences and those connected with drink, drugs or gambling. The suggestion is that the immediate and direct satisfaction derived from such offences as these outweighs any more distant prospect of punishment. By contrast, offences like theft or breaches of traffic rgulations are seen not as an end in themselves but merely instrumental in procuring some other advantage. The potential offender is more likely to calculate, to weigh up the risks, and if he feels they are too great, to get what he wants by some other means: pay for the article he covets or find a spot where he can park his car legally.

There may be some element of truth in this. Even the presence of a policeman may be insufficient to deter an enraged youth or a drunken man from hitting out at someone he thinks has insulted him. A notice in a store that detectives are present and that it is the policy of the management to prosecute anyone caught stealing will make people think twice before popping the odd article into their shopping bags. But it is still an oversimplification. Crimes, as legally defined, are not homogeneous. Violence can be used as a calculated means to gain. Theft can be a direct expression of emotional need, whether the urge to acquire, to outwit others, to impress them, or to attract their attention. And it is not true that the majority of sexual or violent offenders who are caught and punished repeat their crimes: they are, if anything less likely to do so than those convicted of theft. Abnormal sexual impulses do not imply inability to control them in response to ordinary deterrents. It would seem, therefore, that the impulse to immediate gratification can be restrained. It could, indeed, be argued that it is the calculated crime that is least likely to be checked. The would-be thief will simply go to another shop, where the management are more trusting, the staff less observant, the policy less inflexible. My old Professor, Enrico Ferri, used to show his classes an Italian banknote, on which was inscribed the warning that anybody forging one would be sent to prison. Counterfeiters copied the warning, yet they went on forging.

Offences linked with misuse of drink or drugs, with vagrancy, with illegal gambling or prostitution, present different problems. Setting aside the organisers of rackets, who are professional criminals in it for the money, the offences are mostly minor ones, for which heavy penalties are inappropriate. Even when severity has been tried, as it was against the vagrants flogged or transported in Elizabethan and Georgian England, or as it is against the drug addicts imprisoned for years in modern America, its effects have been negligible. Few of the offenders have any stake in respectable society: they have little to lose. After all, they are not deterred by risks to health, jobs, families. Against such a background the whole paraphernalia of the criminal law is water off a duck's back. What they need is not so much deterrence as incentives—an alternative way of life which is within their capabilities yet offers them some hope of satisfaction.

To come back to the point from which we started: for every kind of offence there is one group of people to whom it is almost unthinkable, another to whom it is almost irresistible, and a third who might indulge in it if the temptation got too great or the risk too small. Economists talk about marginal producers—those who will come in when the profits are high and the risks are low and who will be the first to drop out if profits fall or risks rise.

Perhaps we have also marginal offenders, and perhaps it is these who are the main target of general deterrence. At the same time we cannot afford to neglect the big boys of crime: it is they who most expand their operations when they see the profits soar and the risks decline.

Assessing Deterrence

Elementary economic commonsense seems to indicate that when prices are raised demand will fall off, when they are lowered it will increase. Elementary penological commonsense seems to tell us that if punishment is made more severe crime will be reduced, if it is made milder crime will increase. But in criminology, as in economics, what really happens is far less simple.

One of the commonest fallacies is to draw direct analogies from childhood experience of punishment in the family or at school. "I was thrashed when I was a boy and look how much good it did me". But that reflects a picture of the family as a small, close-knit unit bound together by affection and mutual responsibility, in which children can look to the parents for consistent guidance and control. The same can be said, on a larger scale, of relations within a school, at least the kind of school to which those who draw these analogies look back. Both in home and school, punishment could be expected to follow close on the heels of crime, in terms both of certainty and celerity. The situation is very different from the remote, impersonal threats of punishment held out in the modern state. And even for a child, consistency, rather than mildness or severity, is the most important element in discipline. Consistency is the last thing that can be ensured in a society where less than one crime in every four or five is brought home to its perpetrator.

There are very similar objections to attempts to extrapolate from anthropological studies of much simpler societies, or from experience with discipline in the closed environment of the armed forces. In either case there is a homogeneity of values, a predictability, a close solidarity, a direct mutual dependence, an inescapable mutual knowledge and influence, which are almost wholly lacking in a large modern state. In a small, simple and stable society there is at once more support for the law and less need of its deterrent sanctions. In a large, complicated, fragmented and changing society, where other restraints are loosened, where minorities of all kinds are emphasising their differences and asserting their rights, where personal values are uncertain and conflicting, these sanctions are hard to enforce and even when enforced their impact is weakened. In the absence of solidarity, no system of punishments will preserve even the discipline of an army, as has been demonstrated in several recent military conflicts.

Experiments with animals, though they can provide various intriguing clues, are even more remote from the social situation of human beings. Their virtue lies precisely in the ability of the experimenter to exclude everything except the simple choice he wishes his rat to make, the reward which is to encourage a course of action, or the electric shock which is to discourage it. The complications of social life, with its manifold choices and pressures, the convolutions of human conscience and emotions, the uncertainties of human reward and punishment, are remote from all this. Whilst it is true that simplification is essential to answer certain basic questions, it is false to assume that the answers produced can be transposed direct to other and far more intricate organisms and environments.

All this gives some indication of why it is so hard to give straight answers to what seems the simple question "Can we deter people from crime?" The criminal law is only part of a complicated web of deterrence: how are we to distinguish the strength of its various strands, how separate the influence, if any, of penal sanctions from all the manifold and changing social factors in crime? We cannot hold constant other social conditions and influences whilst we measure the effects of a change in the criminal law or its enforcement. We should be hard put to it even to separate these two from each other. Alterations in the methods or accuracy of police recording alone can falsify all our calculations. Nor do we escape the difficulty by comparing two countries with different penal policies: even if their social arrangements are broadly similar we cannot rule out all relevant differences. To experiment simply in order to discover the effects on deterrence of a change in the law, the penalties or the enforcement is rarely possible for political and moral reasons. Where possible at all it is possible only on a limited scale and with minor offences, especially minor traffic offences. These are, in many ways, unrepresentative of crime as a whole.

The next best thing is to seize upon changes made by governments in more serious areas of crime and punishment, to examine the situation before and after, and to try to measure any alteration in the level of the crime concerned. But sudden increases in the severity of the law, or the imposition of exemplary sentences, tend to occur in response to an unusual kind of offence or an exceptional crime wave. When that is so the statistical probabilities are that it will soon level off in any case. Decreases in severity, on the other hand, tend to follow periods when the law has been enforced very leniently, or scarcely enforced at all. Again this reduces the value of any comparison.

In spite of all these obstacles, many attempts have been made to ascertain whether stepping up the penalties reduces lawbreaking. Answers have been sought through most of the range of offences and punishments: fines as a deterrent against illegal parking, mandatory disqualification to prevent drunken driving, longer prison sentences as a threat to hold over persistent offenders, corporal punishment to terrify the violent, and capital punishment as the ultimate safeguard of innocent lives. The short answer seems to be that the upgrading of severity produces most effect at the bottom of the scale, among the minor offences and the financial penalties, least at the top, among the brutal crimes and violent punishments.

University lecturers have been deterred from leaving their cars at forbidden sites by the simple device of fixing and exacting substantial fines: when the fines have been dropped their cars have quickly reappeared right under the notices prohibiting parking. In Connecticut in the fifties the penalties for traffic offences were made much stiffer because accidents had reached a record peak. The sharp drop that followed was hailed as evidence of the success of these sterner measures. But later and closer examination revealed that comparable fluctuations had taken place in the past, without any change in the law, and that neighbouring states had experienced similar trends, again without altering their penal policies. The most that could be said for the Connecticut measures—and it is certainly something—was that the figures did not begin to climb again. Too often, even if the shock of severer penalties produces an apparent reduction, it proves to be no more than a flash in the pan. People quickly learn to live with the new level of penal severity, just as they learn to live with the risks of war or of travel.

Traditional crime and persistent offenders offer still less evidence of the superior deterrent effects of severity. It is true that habitual criminals will often say that the knowledge that their records qualify them for extended sentences will keep them out of crime in future. But it very rarely does. Experience with suspended sentences tells the same tale: deterrence could hardly be made more explicit than through a prison sentence hanging over the head of an offender. Yet the prisons have been crowded with men who have committed fresh crimes even under this threat. Nor is there any evidence that longer terms in prison are more effective in checking recidivism than sentences of under a year. And mandatory minimum terms, whether of life or a long span of years, have the same disadvantages as the death penalty: because they are too rigid they are often evaded and what they gain in severity they lose in certainty.

Just before the second world war, two social scientists, Georg Rusche and Otto Kirchheimer, analysed the penal policies of the main European countries since the beginning of the century to see what effect they had had upon the numbers convicted of crime. Though the statistics they used are open to criticism, there was a remarkable consistency in their conclusion that reductions in criminal sanctions had, in the long run, no detectable effect on the level of crime. Convictions had been falling at times when penalties had become more lenient, rising at times when they had been made more severe and fluctuating when there was no change in the severity of the law at all.

The Special Case of Corporal Punishment

In 1861 the English Parliament accepted the principle that whipping was not an appropriate punishment for adult offenders, though they still thought it was all right for boys. But the very next year there was a transient outbreak of garotting in London: robbers attempting to throttle their victims from behind. As a result, flogging was reintroduced for robbery with violence and remained available for that offence right up to the abolition of corporal punishment in 1948. As it happened, garotting, and the wave of alarm it stirred up, had already subsided before the Bill had come before Parliament. The Home Secretary of the day characterised it as "panic legislation after the panic had subsided". It produced no reduction in robberies with violence: there were even more of them in 1865 and 1866 than there had been in 1862 before the Act was passed.

Nevertheless, the legend lingered on that the "Garotter's Act" had put an end to that particular kind of attack, and time after time in the years that followed spectacular outbreaks of violence were followed by demands that the courts should make more use of their powers to order flogging, and claims that when they did so the violence had been checked. It happened with the 'High-Rip gangs' in Liverpool in the eighteen-eighties, when sentences of flogging were imposed by Mr. Justice Day. It happened in Cardiff in 1908, when flogging was ordered for robbers who had attacked drunken sailors. But in neither case was there any clear connection between the resort to corporal punishment and a reduction in the crime. In Liverpool there were more violent robberies in the early eighteen-nineties, after a prolonged trial of flogging, than there had been in a similar period before it was put into action: 176 in the years 1887–9, 198 in the years 1892–4. In Cardiff the sentences had no immediate effect: it was only after a year's delay, during which no one was flogged, that the attacks began to decline. Experience in Manchester in the nineteen-thirties, when corporal punishment was imposed for living on the immoral earnings of prostitutes, was very similar. In the year 1932 the Recorder sentenced two men to be flogged. But there was no reduction in the offence in 1933 and a rise in 1934. It was only in 1936 that the offence reverted to its 'normal' level.

A longer view strengthens the conclusion that the special deterrent potency of corporal punishment is a myth. In the early years of this century, before the first world war, the proportion of offenders subjected to corporal punishment in England and Wales was drastically reduced: from eleven per cent of those convicted of robbery with violence in the years 1898–1903 to only two and a half per cent of those convicted in 1909–1913. At the same time the average numbers so convicted fell from nearly two hundred to one hundred and twenty-seven. Between the wars, on the contrary, the proportionate use of corporal punishment more than doubled, from twenty to as much as forty-five per cent—nearly half of all those convicted. Yet the crime became more prevalent. It was against this background that the members of the Departmental Committee on Corporal Punishment recommended abolition.

It could still be possible, of course, that although this kind of penalty had no unique and demonstrable value as a general deterrent it was the most effective way of preventing those who actually suffered it from indulging in further violence. The Departmental Committee, and later the Advisory Council on the Treatment of Offenders, combed carefully through the subsequent records of men who had been flogged or birched in all the years from 1921 to 1947, to see whether there was anything to support such an assumption.

In both cases we concluded that there was no evidence that offenders who were flogged as well as being sent to prison did any better than those who simply served terms of imprisonment or penal servitude. If anything, they did slighly worse. This was true whether their subsequent behaviour was gauged in terms of further convictions for robbery with violence, for other serious crime, or for lesser offences involving violence. It could not be explained away by the argument that the men flogged must have had worse previous records than the rest, and were therefore worse risks for the future: it still held good when comparisons were made between those with similar records. And even if it were accepted that, within each group, the courts had reserved corporal punishment for the very worst cases, there was no evidence that they had achieved anything in doing so.

Corporal punishment is by no means a thing of the past. Many countries still rely on it extensively, though reluctant to have the fact publicised. It has, indeed, proved impossible to get any reliable account of its prevalence in the world today.

The Special Case of Capital Punishment

Debate about capital punishment goes back even further, as does the manipulation of statistics by both sides. The controversy, indeed, can be said to have given birth to the whole mass of research designed to measure the deterrent effectiveness of penal sanctions. On the face of it, the death penalty appears indisputably the strongest deterrent of all: "All that a man has will he give for his life". Yet examination and comparison of trends in capital offences throw doubt, to say the least, upon the claim that it has a unique deterrent value. At first sight it would seem comparatively simple to measure trends in murders before and after the abolition or reintroduction of the death penalty in a particular country. But experience has shown that it is as difficult to measure the effectiveness of this supreme punishment as of any other. In certain respects, indeed, it is even more difficult.

First, there are the problems of definition. Murder does not always mean the same thing. It may be redefined, perhaps narrowed. The dividing line between murder and wilful homicide may be shifted. Distinctions may be introduced between murders that are capital and non-capital, or first and second degree. There have been examples of all these in England in the present century: infanticide ceased to be classed as murder; the concept of diminished responsibility allowed certain killings that would formerly have been classed as murder to count as homicide instead; and, prior to the abolition of capital punishment, murder was divided for a time into capital and non-capital categories. Such changes have fostered and complicated debate about the links, if any, between murder rates and abolition of the death penalty. In the United States and elsewhere such calculations have been made even more difficult, since there has been a failure to distinguish, in statistical records, between capital murders, non-capital murders and homicides. Researchers have either ignored this or been compelled to assume, for comparative purposes, that capital murders remain a constant proportion of all homicides, or at least of all wilful homicides. It is no wonder that, here again, it is very hard to get a straight answer.

A second group of difficulties is linked with the fact that in

peaceful and civilised countries murder is a rarity. A handful of
extra cases in a year can produce a dramatic percentage rise, only
to be followed by a less publicised fall a year or two later. Moreover, in interpreting murder trends, with or without the extreme
penalty, it is essential also to bear in mind trends in the rest of
crime, and particularly violent crime, where capital punishment
does not enter into the equation at all. Thus, to take the English
example again, in the sixteen years before the suspension of capital punishment, murders (including cases of diminished responsibility) went up by nearly a half; but indictable crime generally
much more than doubled and violent offences more than
quadrupled.

Finally there are complications arising from the severe and
irrevocable nature of the penalty. Its very existence makes juries
reluctant to convict. Its abolition, by reducing this reluctance,
may thus raise the rate of convictions for murder even if no more
murders are being committed. Furthermore, even where the
penalty is available and pronounced, it is, of all penalties, least
likely to be enforced. Abolition may thus represent less a change
in policy than a recognition of existing practice.

For all these reasons any apparent connection between the existence or application of the death penalty and the level of murder,
homicide or violence needs to be examined with the greatest
caution.

In 1949 and 1950 the British Royal Commission on Capital
Punishment, of which I was a member, examined in meticulous
detail all the available evidence about the effects of abolishing or
restoring the death penalty in various countries. After the most
tedious consideration we concluded that there was no clear evidence in any of the countries that abolition had led to an increase
in the homicide rate, or that its re-introduction had led to a fall.
We warned that the deterrent value of punishment in general was
liable to be exaggerated, and the effect of capital punishment
especially so, because of its drastic and sensational character.

In England and Wales capital punishment was suspended for
five years from 1965 and abolished at the end of 1969. Any attempt to judge the effect of abolition is complicated by the question of whether we are to take, as the basis for comparison, the
numbers of offences initially recorded by the police as murders,
or the numbers resulting in an eventual conviction for murder. In

trying to assess the deterrent impact of capital punishment upon a capital crime we need to look at both. The number of offences initially recorded by the police as murder rose in most years between 1961 and 1968 (from 214 to 335): the rises between 1964 and 1966 (242 to 266; 266 to 305) were among the sharper of the period. There was a decrease in 1969 (to 309), but thereafter there was a steady rise to 520 in 1974. However in most years a majority of these offences are found by a court to be manslaughter, and only a small proportion are found by a court to be murder. If figures of offences found by a court to be murder are considered an upward movement is found to have begun between 1963 (38, which was also the figure for 1961 and 1962) and 1964 (47)—that is, while capital punishment was still in force—and to have continued in most subsequent years until 1970 (91); thereafter the figures were a little lower (85; 88; 82) until 1974 (148, with 7 cases still pending: this was a year in which the figures were inflated by murders associated with the Irish disturbances). The number of persons convicted of murder has tended to increase over the period 1961 to 1975 (from 51 to 107), but there have been substantial fluctuations, some of them downward, from year to year. Taking these sets of figures as a whole, there is no evidence of a sharp change in pattern around 1965 sufficient to demonstrate that capital punishment was a uniquely powerful deterrent.

Nor is there any evidence that the restoration of capital punishment after a period without it serves to check the rate of murder. In 1961 there were four particularly shocking murders in the State of Delaware: a widow of nearly ninety was beaten, raped and stabbed, an elderly couple were shot on their farm, a woman at work in her kitchen was killed by a bullet through the window. As a result of public feeling, the death penalty, abolished four years earlier, was restored. Looking back now, it is clear that the rate of murders was lower during the years of abolition than either before or since. More intensive studies, in Philadelphia and Chicago, of what happens when the death penalty is actually enforced produce the same picture in miniature. Neither the publicity at the time of the offence and trial, nor the execution itself, appeared to have any influence upon the fluctuations of murder in these cities. An international survey carried out in 1962 by the United Nations Department of Social and

Economic Affairs confirmed the general opinion that neither suspension nor abolition of the death penalty had any immediate
effect in increasing the incidence of that crime. Countries which
had abolished it, like Germany, Austria, Scandinavia, the Netherlands and Denmark as well as several Latin American states, reported no ill effects. Finland, indeed, reported a steady decrease
in murders.

None of this implies, of course, that to drop the death penalty
is a recipe for reducing murder, though if the findings went in
the other direction there is little doubt that many would infer
clear cause and effect. But there is no shred of evidence that the
threat of execution is a more effective deterrent than the threat
of life imprisonment which usually takes its place. That has been
the case even though, because of the chances of licence, parole or
amnesty, life imprisonment has seldom meant what it said.

Murder rates in fact have been rising far less than those of
other offences. Where they do rise sharply, as they did in France
during the Algerian crisis or are doing in the United States or
Northern Ireland now, it is for reasons that go much deeper than
the nature of the penalty.

This is borne out by comparing murder rates and the availability of the death penalty in different countries and states. The
exercise is hampered by many differences: in legal definition, in
the practices of prosecutors and courts, and above all in political,
social and economic conditions. Even within the United States
there are wide differences in homicide rates just as there are wide
social differences between states. But when we compare states
which are similar in other respects, but differ in whether they
have retained or dispensed with the death penalty, we find that
they have very similar rates of homicide. Again, it is sometimes
assumed that only states or countries with comparatively few
murders feel able to abolish capital punishment, and that this is
why they feel the effects so little. But that too is contradicted by
the American experience: some states with high murder rates
have abolished the death penalty, some with low murder rates
have retained it. Yet it can hardly be retention that has kept the
rates low. There are states holding on to capital punishment with
much higher rates than others which have relinquished it.

We are brought back to the conclusion that it is social and
cultural conditions that determine both the murder rate and the

penal response. Incidentally they also seem to determine the frequency of lynching, thus contradicting yet another legend. It is sometimes suggested that, in the absence of the death penalty, outraged citizens are likelier to take the law into their own hands. But as executions have declined in the United States so have lynchings, and the Southern States, where most of them occur, all have the death penalty as well.

Though there may be a dearth of evidence to prove that capital punishment is a unique safeguard against murder in general, it is often claimed, nevertheless, that it has a special importance as a check on the calculating professional criminal, and as a protection for people like policemen and prison officers who may have to deal with desperate offenders who have nothing to lose but their lives. The professional criminal, it is argued, calculates his risks. If he is not deterred from bank robbery, he may at least be deterred from carrying arms, or from adding murder to the rest of his crimes simply to escape detection or arrest. Again, on the Royal Commission on Capital Punishment we found no evidence from other countries that abolition had led to such consequences, though we conceded that, if capital punishment had any unique value as a deterrent it was here that its effect would be chiefly felt and here that its value to the community would be greatest. It is true that, since the death penalty was first suspended then abolished, violent and professional crimes have been going up in Britain. But they were already going up before. In particular some of the most notorious and best-remembered murders of policemen took place whilst hanging was still the penalty.

In some parts of the United States the death penalty has been retained only for the murder of a prison officer by a prisoner already serving a life sentence. In England the Royal Commission gave exhaustive consideration to the possibility of making similar distinctions, but we were driven to the conclusion that they would only produce anomalies. The murder of a policeman would be capital: that of a Prime Minister would not. In any case is it logical that the police or prison staffs should have more protection than other citizens against the villain who has nothing to lose? However great the indignation, especially when an unarmed policeman is shot by an armed robber, the fact remains that the policeman, like the coalminer or lorry driver, undertakes certain risks as an inevitable part of his job. And there is no more empirical evidence that the threat of the death penalty gives greater protection to police and prison officers than that it gives to anyone else. It has been demonstrated, for example, that they are just as likely to be murdered in those American States which have dropped capital punishment as in those that have retained it.

A feature of this whole controversy about the effectiveness and the justification of the death penalty as a deterrent is the fact that, in civilised countries, it is a punishment that has seldom, if ever, been enforced with anything like full vigour. It is possible to argue that, if it were, its relative power as a deterrent would rise sharply. However, even that may be questioned when it is recalled that murders tend to be committed either impulsively by people acting under very strong emotional pressure, or more cold bloodedly by people who have calculated their chances and covered their tracks. Virtual certainty of death, if discovered, has repeatedly failed to deter members of underground movements or those who conceal them in war.

Nevertheless it would be wrong to underestimate the hold capital punishment retains upon the public imagination. It is still seen as an indispensable adjunct to the maintenance of law and order and it offers to satisfy the deeply felt need for retribution and expiation in relation to certain atrocious crimes. Statistical evidence has, in the past, been heavily relied upon, often distorted, by abolitionists and retentionists alike. It has proved, however, to be far too tenuous, too complicated and ambiguous, to make much impact upon public opinion. Utilitarian arguments for the restoration of capital punishment have been shifting from deterrence to the more tangible problems. How can society be secured against the repeated onslaughts of known and ruthless criminals in times when their numbers and their crimes are multiplying? Can they and must they be held in confinement virtually for the rest of their lives? If so, what will be the impact on the prisons? And how are we to combat the threat of intervention by terrorists to secure the release of their unlucky comrades?

When we look at the world as a whole, the careful researches, the anxious interpretations, going on in the liberal countries cannot fail to appear esoteric. Moreover, according to the latest enquiry carried out under the auspices of the United Nations, capital punishment figures in the legislation of no less than a hundred and twenty-five member states. In many it is applicable not only

to murder but to a number of other offences as well, and in some
it is being enforced by way of public executions. In contrast, the
countries which have abjured its use are a mere handful. The
Supreme Court of the United States, in a famous judgment, condemned it as a 'cruel and unusual punishment'. But in the world
perspective it is still a very usual punishment indeed.

Yet consider the cost, in a peaceful civilised society, of enforcement sufficiently rigorous to leave the killer in no doubt of
his fate if convicted. We should have to carry out the death
penalty consistently, over a long period and on a large scale,
cutting down ruthlessly the considerations of age, diminished
responsibility, circumstances, that normally provide loopholes for
commutation. We should have a hanging every three days in
England, the electric chair in use daily in Detroit. The extensive
and rigorous application of the death penalty simply would not
work in societies such as ours.

The Level of Enforcement

If severity of punishment is least effective in precisely those
spheres where we are most anxious to deter, is there any hope
that we can, instead, keep down crime by stepping up police
vigilance and increasing the risk of being caught? Certainty of
detection and punishment, as envisaged by idealists like Beccaria
and even by such hard-headed realists of the nineteenth century
as Sir Edwin Chadwick and Sir Robert Peel, is never likely to be
achieved. But to increase police vigilance, to give them the resources to follow through more of the offences that come to their
notice, is not beyond the bounds of possibility. The question is
whether it is worth it, whether it can really have any effect on
the level of crime.

In 1954 the police in North-west Manhattan, one of the high-
crime areas of New York, were strongly reinforced for a period
of four months. This period was then compared with the corresponding four months in 1953. Variations in murder and rape
were no greater than could be accounted for by normal fluctuations. But assaults dropped from 185 to 132, big thefts went right

down from 153 to 46, car thefts from 78 to 24. On the other hand arrests for some offences—the kinds where action depends almost entirely upon police initiative—showed a marked increase: 186 arrests for drug offences as against 78 the year before; 170 for gambling as against 125; and 372 juvenile cases as against 135.

The New York experience illustrates the special difficulty of assessing the effects of stepping up police activity. A larger proportion of offences will be noticed, dealt with, recorded, and that in itself will produce an apparent rise. But the more serious the offence, the greater the likelihood that the victim himself will lodge a complaint, and therefore the less the impact of variations in recording. And this New York experiment does bear out the hypothesis that greater vigilance can bring about a reduction of the middle range of offences, whether of violence or dishonesty. It may well be that saturating the streets of Manhattan with police merely encouraged the local criminals to transfer their attention to other parts of New York. The London suburbs complained of the same thing when the Metropolitan Police were first established in the eighteen-thirties. But that makes a case for better policing everywhere rather than giving up the attempt to deter by watchfulness.

Not unexpectedly, experiments in strengthening traffic police have also produced measurable improvements. The introduction of the breathalyser, giving people the impression that they might at any time be stopped and checked by the police, was more effective than the threat of disqualification in reducing drunken driving in England. Once it was realised that the chance of being caught was no higher, even though the chance of conviction had been increased, drunken driving was again on the upgrade. Similarly an increase in police patrols on a busy stretch of highway in California was followed by a reduction both in accidents and in driving offences, as counted by independent observers.

Some of the strongest evidence of the importance of police in the day-to-day prevention of crime is derived from occasions when their vigilance has been suddenly withdrawn. The most prolonged instances of such a situation have been part and parcel of revolution or war. The French Revolution undoubtedly loosened the restraints against crime of all kinds—what is surprising is that so many people continued to go normally about their business and lead honest lives. Shorter but sharper was the episode during the second world war when the Gestapo arrested and removed the entire Danish police force, leaving only a makeshift, unarmed watch to enforce the ordinary criminal law. Before the war no more than ten robberies had been recorded in Copenhagen in a whole year. With war conditions the rate rose to ten a month by 1943. But after the removal of the police in 1944 the crime multiplied tenfold, and that in spite of the severe penalties imposed on the few who were caught.

Of course, in defeat and revolution the relaxation of police vigilance is only one of the factors that may be expected to stimulate crime. The occasional police strike, though much briefer, offers a clearer opportunity to see the effect of reducing the risk of detection alone. In Liverpool in 1919 nearly half the police went on strike from midnight on July 31st. Looting of shops began the following evening and continued for some days. About four hundred shops were raided and there were many assaults upon those constables who remained on duty. In that year there were less than seven thousand cases of shopbreaking in the whole of England and Wales: over a thousand occurred in Liverpool. Again, in 1970, when the police went on strike in Montreal, the city came perilously close to being taken over by criminal elements. This is the sort of story that will always tend to repeat itself.

Deterrence is seldom given full rein in any society. When it is, it can be very effective. But there are several prerequisites. It must be sustained, over-shadowing and weighing down the whole life of the community. And it must be consistent with the rest of that life. It calls for a homogeneous society, with homogeneous values. Attempts to impose it upon a reluctant or subjugated community may have short-term and limited success, but some of the people will resist all the time, nearly all of them some of the time. We may recall Durkheim's warning that a society which would wholly repress crime would, in the process, have to repress all initiative, all nonconformity, all adaptation to change. Whenever we contemplate more drastic deterrence for criminals we need to consider also what wider social consequences it will bring with it for the rest of us.

Deterrence, seen as a threat based on power, has been an unpopular theme amongst those who claim to be progressive. Unfortunately we cannot dispense with it as a defence against crime.

It still has a vital protective function in society. The more keenly we appreciate the risks that it may be applied in the wrong directions or the wrong intensity, the stronger the case for studying its working and ramifications. There is cause to be thankful that no state has yet found a recipe for completely successful deterrence. If it had human liberty would be at an end.

Part IV

ENFORCING
THE LAW

CHAPTER

7

The Police: Protectors or Oppressors?

The Problem

Like most people I would rather live in a world with no police—provided, of course, that I could rely on everyone else to behave in a reasonable and orderly way, with due respect for my possessions, my life, my rights and my safety. Since no such ideal state exists, I accept that we must have police, whatever the dangers implicit in their powers. "Corrupt or not," remarked Judge Knapp at the height of his exposures of police abuses in New York, "the police respond to duty. If anything happened where I was in trouble, I would have absolutely no hesitancy in dialling 911 and being certain that the men would react." Even dwellers in American slums have complained as much of police neglect as of police abuses, have demanded more police attention and protection rather than to be let alone. In the last resort, whatever their shortcomings, the police are regarded as the upholders of the law, controllers of disorder, protectors of the public, and that goes for the poorest and most defenceless at least as much as for the prosperous.

To carry out these tasks they must have both power and discretion. The temptations to abuse power, to twist discretion into improper discrimination, are always there. Whether in picking out and questioning of suspects, in control of crowds, or in day-to-day interventions in private quarrels, there is always the risk of improper bias or improper means. The major problems of policing stem from its very nature.

Police power, in itself, sets them apart from the public. Only very rarely, as when an English policeman is shot in the course of his duties, can they expect evidence of sympathy or liking. But it is remarkable that, according to surveys in the nineteen-fifties and sixties, they have been able to command confidence from the majority of citizens at least in parts of the world where law enforcement has followed the Anglo-Saxon model. In England and the States, in Canada, Australia and New Zealand, more recently in Israel, the verdict then was much the same. Seven or eight out of every ten citizens questioned, though aware that the police were occasionally venal, occasionally brutal, expressed general respect for them. In England four out of ten had called on them for help in very varied emergencies over the preceding ten years, and only four in a hundred had been disappointed by the response. Even amongst those groups least likely to approve of the police—the young, the very poor, racial and other sensitive minorities—the level of approval seldom dropped below five out of ten. Only a tiny minority expressed outright disapproval.

Yet those same surveys showed seven out of ten of the police, on the other side of the gap, very pessimistic about the public view of them. One obvious reason is that they deal mostly with the people likeliest to distrust them: the young, the recidivists, minorities at odds with the way of life of the majority or with the outlook of those in power. Vicious circles of mutual suspicion and hostility can develop only too easily between such groups and the police, and are then very difficult to break.

Part of the problem lies in the distribution of known crime. If some of these groups in fact produce more than their share, the police will inevitably turn first to them when they are trying to find out who is responsible for an offence. The people concerned resent it, become even more hostile, perhaps even more involved in crime as a result.

Suspicions of police partiality, however, derive also from the existence of the dark figure of crime. There is the problem of under-enforcement of the law against the respectable and well-to-do. Middle-class young hoodlums have traditionally been allowed to get away with wanton damage to property if they or their parents could pay for it. That may no longer be the general rule in England but fairly recent research found it still very much the rule for a group of such privileged delinquents in California. Their elders, the white-collar criminals of business, are even more frequently immune. The police have little access to their strong-holds, few officers trained to find their way through the laby-rinths of frauds, scant resources for the lengthy investigations needed to penetrate to the heart of conspiracies. More successful enforcement of the law in these privileged circles could lessen the brooding sense of injustice felt by many people, the feeling that the police are respecters of persons, applying one law for the rich, another for the poor.

Then there are the ways in which police discretion and use of their powers are distorted by the prejudices, fears and suspicions they share with the majority of their fellow-citizens. People liv-ing in so-called ghettoes of Blacks and Puerto Ricans in American cities are twice as likely as the more prosperous to report that they believe the police to be disrespectful, brutal and corrupt. They complain of violence in manner and action, the overriding of people's rights, gross discrimination against the black and the poor.

There is an absence of systematic and rigorous evidence as to whether the police do discriminate in racial terms in the handling of criminal suspects. Such evidence would be hard to get, since the blacks who come to the attention of the police are commonly also poor, and since there is so much mutual suspicion and tension between the police and poor black communities. Yet observation, and many case records, suggest that bias exists and there is a feeling that black suspects are likely to be taken into custody on less evidence than whites.

On top of that there are problems of police that derive from the wider problems that society at large has ignored or failed to solve. Take, for instance, the question of minority values and ways of life: how far should they be accepted and preserved, how far should they be replaced or absorbed into the majority culture? A youth from one American city slum could not secure parole from prison because of his stubborn refusal to admit that his attack on his girl friend had amounted to rape. In his eyes he had merely followed the normal courting pattern. Should the police and, for that matter, the courts, accept all but the most serious assaults and fights as an inevitable part of the local culture, wherever possible avoid bringing formal charges? Or would that mean they are acquiescing in lower standards, perpetuating the gulf between these people and the rest of the nation, reducing still further the chances that they and their children will ever break free of violence, denying them the protection that others take for granted? The balance between a proper respect for local traditions and ways of life and for the long-term interests of minorities is a delicate one. The police, often ill-prepared, are assigned the job of maintaining it in far from delicate conditions.

Or, yet again, there are the problems posed by the sheer poverty of certain classes and areas, the weakness of informal social restraints, guidance or leadership, the paucity of social services, the squalor of housing and family life, all pushing the police towards formal action by way of criminal proceedings rather than diversion to some alternative kind of control.

But do all the problems of police stem from the social conditions in which they work? What of suspicions that it is only those who want to dominate others who choose policing as a career in the first place, that police are, at heart, born bullies, rigid, conventional, authoritarian? Whilst one American researcher has contended that they come mainly from the working class and have therefore a penchant for tough language, physical force and prejudice against minorities, another has categorised them as essentially of the lower middle class, preoccupied with order, cleanliness, thrift, punctuality and conventionality of behaviour and dress. The author of a classic monograph on the police in Britain and the States, on the other hand, found no evidence of a cut and dried police personality at all: in his experience they ranged "from the tender-minded to the flinty, from the clown to the mandarin."

It hardly needs research to assure us that most police are conservative in their attitudes and would deny that citizens have a right to disobey laws, even if they consider them unjust. It is almost a matter of definition, and is as true when they are defending the values of the Soviet Union or of Chairman Mao against deviance as when they are enforcing the law, or defending property, in a western democracy. The radical or extreme liberal who strays into a police force is unlikely to survive there for long, unless he is shunted into some kind of specialised 'community relations.' However, there have been the questionnaires and in-

terviews designed to test more directly police motives for joining. Less than half, in one American survey, claimed that a career
in police work had been a long cherished ambition: for the rest it
was simply a regular job with a reasonable status. If a further
motive had to be produced it was likelier to be service to the
public than a desire to dominate them. Who, after all, is going to
admit to that? A firm of management consultants, asked to probe
the motivation of a hundred senior police in Chicago, came up
with the reassuring news that they were 'not tough, cynical or
hard-bitten, but decent, kindly men, intent on doing their jobs,
with marked social service values.' How does that square with the
behaviour of the Chicago force during the Democratic National
Convention of 1969, when they subjected both demonstrators
and bystanders to indiscriminate violence, amounting to a police
riot?

If it is not personality that sets police apart from, sometimes
against, their fellow citizens, it could be their training and experience. The most depressing feature of the former is its narrowness
and isolation. I have seen police academies in many cities and
there is the same coaching in elementary police law, the training
in the use of guns, all carried out by police officers. An occasional
smattering of sociology is beginning to creep in, but there is no
exposure to the views of the long-haired or the drug addict, no
thorough analysis of the dilemmas and decisions that will be met
on the streets. How can the problems of police discretion in law-
enforcement be clearly discussed as long as its very existence is
officially denied?

Experience on the job has been pictured by some researchers as
a depressing 'cop's progress' through a career of deepening disillusionment. They trace the shocks encountered by the new recruit when he first goes on the beat, the growing cynicism
produced by public hostility, the hardening compromises and
exigencies of the job, entrapment in illegal short-cuts, perhaps
even in a group of police who demand conformity in corruption
or participate in outright crime. They distinguish the pseudo-
cynicism of the recruit, anxious to prove he is no fool, the ro-
mantic cynicism of the disappointed idealist, the aggressive cynicism produced by ten years on the beat and finally the resigned,
apathetic cynicism which takes over as retirement approaches.
On the other side we have a sample of New York police tested

when they first entered the service, and again after ten years on the job, to see what they thought the most important qualities in their work. The recruits stressed appearance, alertness, reliability, courage; the old hands stressed honesty, dedication, emotional maturity, compassion, good temper. Again one recalls the recent revelations about New York police: how do they chime with these ideals of honesty and dedication? Or should we just accept that the police, like the rest of us, are a pretty mixed and mixed up lot? As for professional progression through deepening degrees of cynicism, or a deepening emotional maturity, the stages could equally be found in many other professions I could name.

The kind of force in which a man serves will obviously make a difference, limiting or extending his scope for discretion, whether for good or ill. So will the nature of his duties in the force: the attitudes developed by detectives are different from those of men on the beat, or in riot squads, or in work with juveniles.

Certain qualities, however, are inevitably developed by service in the police: a sense of danger; watchfulness; suspicion of the unusual; an attitude of authority; a measure of cynicism; a solidarity and loyalty towards colleagues; secrecy; jealousy of outside intervention. The police would not get far without most of those qualities. Yet they do serve to widen the gap between them and the public. And because police powers are great, their possible impact on individual lives extensive, exaggerations, misdirection or distortions of these essential qualities are particularly dangerous.

The otherwise optimistic Royal Commission on the Police foresaw a future deterioration in relations between police and public in England. But another reliable survey in 1973 showed that the proportion of people expressing a great deal of confidence in the police had been maintained at seventy per cent (sixty-five per cent even amongst the youngest, least trustful group). This placed them well ahead of other major national institutions, including law courts, education, the church, medicine and trade unions.

Nevertheless, the trends of modern societies have tended to aggravate rather than alleviate the problems of the police. They have not been able, except sporadically here and there, to check the rise in crime or the spread of disorder. Minorities of various kinds are becoming ever more assertive, even violent, in pressing their claims, and they often choose the police as obvious and immediate targets, representatives of what they see as an unjust 'establishment.' If a policeman can be induced to retaliate he can be held up as an example of state brutality. Then there is the growing cynicism about authority, the dwindling stigma attached to law-breaking, the whole 'permissive' ethos, which again makes people less responsive to the police, often leaves the police themselves bewildered about what is, or is not, to be tolerated. Finally, there have been the much publicised scandals of the last few years, often beginning with a whisper, scarcely believed, confined to accusations against one or two officers, then opening out to reveal depths of corruption, breadths of involvement, far beyond what was originally suspected.

The Role of the Police

In the context of rising crime and civil disorder, of conflicts and stresses between individuals and groups throughout society, what is the primary task of the police? Is it to catch major criminals; to control major disorder; or day-to-day public service of advice, conciliation and peace-keeping? We can seek the answer by looking back to see for what purposes police forces have been developed in different parts of the world. Or by asking how much importance the police, the public, governments, attach to each of these functions now.

Historically the purposes and tasks of police forces have always been mixed, the mixture varying with the political and social situation.

For centuries in England there was pride in their very absence; the conviction that no free people could tolerate either a standing army or a professional police. Two things changed all that. One was the nineteenth century zeal for penal reform. If only we could be more certain that felons would be brought to justice much milder penalties would suffice to deter them and England need no longer be the only civilised country to rely on the threat of hanging to keep down petty theft. But at least equally important in securing acceptance of regular police was the threat, real

or imagined, from unruly crowds. Those were the days of Peter-
loo and Reform Bills, of Chartists and Tolpuddle martyrs. When
crowds of working people got together to shout for their rights
the last thing really wanted was to set the soldiers on to them.
Stern old Tories like the Duke of Wellington feared that guards-
men, called upon to quell their compatriots, might forget their
loyalty. Cunning old radicals, like Francis Place, stooped to advise
Sir Robert Peel on how police might avoid direct confrontations,
break up mobs without violence. So the English got their police,
in the last resort, in the interests of moderation: moderation in
penalties for criminals and moderation in control of the disorderly.

The role set before the pioneer metropolitan police of London
was likewise framed in terms of minimum force and maximum
service to the public. The first Commissioners were determined
to allay public suspicion and resentment and to transform public
hostility into appreciation and co-operation. High standards of
behaviour were to be exacted: drunkenness and dalliance were
forbidden; high-handedness in such matters as telling people to
move along was stamped upon; even street pedlars were to be
treated with civility and compassion. The police uniform was to
be civilian rather than military and they were not to carry arms.
Above all the motto was to be the prevention of crime rather
than the mere detection or capture of criminals after the damage
had been done. The police were to make themselves so familiar
with their areas and all those who lived in them that they would
immediately spot anything out of the ordinary and detect indications of crime before they could be carried into effect. And they
were constantly to be warning the honest public of new criminal
devices and the precautions that should be taken to foil them.

This rosy vision of a police so deeply in the confidence of the
public that together they could nip crime and disorder in the bud
was never fulfilled. Policemen did get drunk, were sometimes
bullies, even spies. Gradually and reluctantly a detective department evolved, with plain clothes police taking their place alongside the open, numbered, uniformed bobby. Even later a small
'Special Branch' had to be formed to counter subversion, whether
native or foreign. The dreams that crime might be prevented by
the mere presence of police on the streets, that habitual criminals
could be stamped out by early detection, faded into the long-

drawn, humdrum facts of continued crime, partial police success, with considerable police and public disillusionment. Yet it is still true to say that in England, on the whole, the public rely on the police, expect their prompt help in times of trouble and regard them as servants rather than masters.

In Continental countries like France the roots and roles of the police have been very different. There political considerations have been paramount. With wide and diversified territories, long land frontiers, centuries of invasion and provincial strife, it was inevitable that the central power, if it was to survive at all, should be served by police forces that were armed and that could gather political intelligence from all over the realm. The French police, and the other continental forces modelled upon them, have traditionally been closely linked to territorial armies and have played a major role as political informants. Almost every Continental country maintains a separate riot police, frankly armed as soldiers, frankly on the watch for any stirrings that could be classed as subversive, often provoking the organised violence on the streets that it is its ostensible task to quell. If in England the role of police as servants of the public still dominates, in France and much of the Continent this other picture of the police, as masters of the public, colours the role they adopt, the ways they behave and the ways people react to them.

In the United States police have combined, in a curious way, some aspects of the public service role expected in Britain with the toughness taken for granted on the Continent of Europe. The mixture of populations from which they come and with which they have to deal, especially in the cities, together with the exceptionally high level of crime and violence, combine to produce this duality. At one moment you are looking at what the Philadelphians came to call their 'skull-cracking brigade' of riot police, or at scandals like the police riot in Chicago. At another you are hearing of the thousands of calls from citizens answered every day and night of the year, in the attempt to ameliorate disputes, make peace between quarrelling neighbours, calm down fights on the streets or in cafés—all of it dangerous work, yet all of it a police service taken for granted, at least in some parts of the country.

In the vast areas of the world that have been finding their way towards a sometimes precarious independence and sense of na-

tional identity, the political role of the police has again predominated. Both their colonial inheritance and their present exigencies
have contributed to that. They too have had to cope with long
frontiers, warring local interests, breakaway tribal, racial or religious groups, subversion, both internal and external. On top of
that they have had the massive problems of development, of shifting people's ideas, of transforming traditions, coping with the
influx from country to town, meeting crime on a wholly new
scale. In all this governments have used their police, who have
been expected not only to bring criminals to justice, but also to
keep tabs on opponents of government, and to act as administrators, tax collectors, and armed forces in the face of unrest. Inevitably they appear as masters rather than servants to the general
run of their countrymen.

How are the various functions of police evaluated nowadays?
In the view of most modern policemen it is catching criminals,
especially the violent and the organised, that constitutes the real
core of the job. The elite, in the eyes of both police and public,
are the detective branches, the wide-ranging regional crime
squads. The most advanced modern technology must be engaged
to centralise, sift and transmit information, to overtake the mobile
modern criminal. There must be specialist units to match the
specialisms of crime: the drug squad, the vice squad, the fraud
squad, the experts on thefts of art and antiques, not to mention
the Special Branch. There must be advanced forensic facilities.
There must be centralisation of control, interregional and international co-operation. The police are to be seen as professional
soldiers in a war against crime, to be organised and equipped as
such. The role of the public is to notify anything suspicious, co-
operate in enquiries if required, and keep out of the line of fire.

The crime detection role of the police gains further weight
from the fact that it appears, at least superficially, the easiest to
evaluate in objective terms. It is possible to compare the crimes
recorded and cleared up, the offenders brought to justice and
convicted, from one year to the next, by one force and another.
Detection rates become a touchstone in judging police achievement, whether for forces or individuals. It all seems very convenient and convincing until we recall how much crime goes
unknown or unrecorded, how easy it is to juggle or interpret the
figures. A police force which is not particularly observant, which

earns little trust from the public, which is careless or downright dishonest in recording even the crimes of which it is aware, which does not hesitate to use brutality or bribery to extort confessions, true or false, can put up a better statistical picture, claim a higher proportion of known crimes cleared up, than a force that is thorough in its methods, and keeps honest records.

Undue stress on this part of police functions carries other risks. The more police are preoccupied with the necessity of bringing to book those they regard as dangerous villains, the more impatient they become of legal restraints on their powers and discretion, the more convinced that the end justifies the means. Again, the more the police become mechanised, technical, specialised and centralised, the greater the distance between them and the public: what they gain in expertise and speed they can lose in grass-roots observation and information.

The role of the police in maintaining public order shades only too easily into political partisanship. Yet it is a role they cannot escape. The most liberal government is bound to seek information about people it suspects of preparing to use violence for political ends and to try to forestall them. It is bound to take action to control violence by crowds, however idealistic. And the police are the obvious instruments. But there is a fine line, in terms of strategy, between on the one hand, preventing outright crime and protecting existing institutions from overthrow by force and, on the other, using force to suppress legitimate change or protest. What if the police serve a regime that defines all dissenters as dangerous rebels or that crushes terrorism from one end of the political spectrum, but ignores it from the other? A police force would not be doing its job if it did not try to prevent rioters from building barricades, burning buses, hurling paving stones, but what if it takes sides?

There is also a fine line in terms of tactics: the decision as to whether, when and how, force should be used or threatened to prevent disorder from getting out of hand. A mistake, even in good faith, can stir up suspicion of police bias, fan the riot they are trying to prevent. The success of the police in this field is measured not only in terms of their success in preventing the spread of violence, looting or arson, but in the after-effects on police relations with the areas or groups involved, and on public opinion at large. In these terms a tactical victory may be a long-term strategic defeat, a black mark against them, remembered and embroidered for generations. The 'security' activities of police provoke, at best, uneasiness, at worst, bitter and continuing hostility. As limited as possible, as controlled as possible, as answerable as possible, seems the best set of maxims for police engaged in these perilous duties.

The role of police as public servants and peacekeepers has been the least measured and least regarded of their functions. There is a vagueness, a variety, a lack of formal definitions and formal powers about it. Yet it accounts for many more calls for help, many more encounters between police and public, and much more of the time of the ordinary policeman, than duties stemming directly either from traditional crime or from threats to public security. The police have picked up many jobs simply by being available. Public and governments alike know they are on the streets, or in their stations, all round the clock and call upon them accordingly. In many countries it goes further than that. The more the police are involved in day to day administration the wider their information and control. The continental conception of the role of the police, like the continental conception of public order, is much wider than the British. Lack of other services and other resources has brought many of the developing countries to a similar position: both colonial and national governments have found it convenient to use the police for tasks like tax collecting that fall to civilian officials elsewhere. And throughout the industrialised world police forces have been spending nearly half their time on controlling motorists.

Why not transfer all these time-consuming jobs to traffic wardens, as in New Zealand, to tax collectors, health or social services, and leave the police to concentrate on their more specific tasks? Several answers have been advanced. First, these seemingly borderline duties are much more closely tied up with the prevention and detection of crime than might appear at first sight. The police claim that their powers to direct and stop traffic, to check on licences and insurance, are essential if they are to catch the motorised criminal: and most criminals nowadays, like most other people, travel by car. Organised crime, in particular, depends heavily on a succession of stolen vehicles. The possibilities of violent crime likewise lurk in the family rows, the pub quarrels and so forth. It is true that only a small proportion of them lead to really serious assaults or murders. Yet, looked at the other way round, more than half of all murders and grave assaults occur within the family or among friends or acquaintances. Again, it may be true that the police are not properly qualified to deal with domestic problems, with juvenile delinquents, with drunks or drug addicts and the rest. But at the very least it is they who have the task of immediate action, investigation and, where necessary, first aid. We may provide shelters, consultants, hospital beds, hostels, social workers, drying-out centres. The chances are that it will still be the police who have to get people to them in the first instance and that in doing so they will be helping to maintain the peace and prevent offences.

So perhaps it is a matter of recognising rather than rejecting these social service elements in the role of the police and of making more place for them in training, organisation and liaison. In the last few years there have been signs of this, if only straws in the wind. There has been beat policing, the attempt to recapture, in a small network of city streets, the intimacy, personal knowledge and personal responsibility of a village policeman. In London they have been trying one man permanently on one beat, a small local car to go to the small local emergency, and someone responsible for recording and assessing the trends in that particular neighbourhood. In Harlem they have experimented with a Family Crisis Intervention Unit: a group of police officers specially trained to handle family violence. In Washington there has been a neighbourhood centre, manned round the clock not only by police but by local people employed and trained by welfare and legal aid agencies. Another American proposal was mobile emergency units of social workers and psychiatrists, ready to join the police or to take over from them, in dealing with family disputes or problems. In England there have been attempts to build up regular consultation between police and social services in dealing with children in trouble, and there have been juvenile liaison schemes.

In districts with mixed populations, efforts have been made to enlist more blacks or immigrants in police forces. Half the cities in the United States now have community relations officers. In view of the obvious complexity and magnitude of the task these are no more than tiny beginnings.

The informal, intangible aspects of the work of the police

remain the hardest of all to measure and assess. Where researchers have studied the amount of time spent on the social service
and peace-keeping functions of the police, the numbers of calls
for help received and the responses made, they have demonstrated that they account for far more police effort than is
commonly recognised. It is only where they are missing or
inadequate that they get much notice from press or public. Yet if
they were withdrawn we should slip back to the law of the
jungle in day-to-day life, feel the lack of protection and concern
that tastes so bitter to people in underpoliced areas, whether in
the poorest parts of cities or the poorest parts of the world. The
concept of the police as people you can always turn to in an
emergency is too rare and valuable to be taken for granted.

Not that the role of police as peace-keepers, informal advisers
and controllers escapes the problems of misuse of powers and
discretion, of unsolved social conflicts or of prejudice or rigidity.
A policeman on every corner, knowing everybody's business,
could undoubtedly deter from crime and disorder. But would he
really make people feel more secure? To enforce the law or
prevent crime is indubitably part of the job of the police: to
impose their own ideas of what is moral, proper and permissible,
beyond the limits of the law, is not.

Police Abuses

Corruption

Perhaps the most spectacular revelations of widespread police
corruption in recent years have been those of the Knapp Commission appointed in 1970 to investigate allegations by former
policemen about what had been going on in New York. The
Commission was told that all but two out of seventy-five police
working in one sector of Brooklyn regularly took small bribes
from gamblers and supermarkets for turning a blind eye to lawbreaking. Plain clothes men in Harlem extracted well over a
thousand dollars a month from gambling establishments they
were supposed to be watching. In one instance eighty thousand

dollars had been exacted as the price of non-enforcement of the narcotics laws. Detectives collected bribes which sometimes ran into thousands of dollars in the course of other criminal investigations. And the uniformed patrolmen generously supplemented their income by small payments from the myriads of people continually breaking laws and regulations: from building sites, bars, grocers, gamblers, prostitutes, madams, and motorists. There was even widespread bribing of policemen by policemen, just to speed things up or to put them on duties they coveted.

Corruption as pervasive as this cannot be written off, as Commissioner Murphy tried to write it off in 1971, as the work of 'one or another traitor to uniform.' Nor has it been peculiar to New York. Most American city police forces were created in the latter part of the nineteenth century when political corruption was at its height, and they were deeply involved in that corruption. They have at the same time been highly vulnerable to organised criminal rackets, whether in alcohol or prostitution, gambling or drugs. The case has been put on record of a city where the police chief was on the payroll of the illegal gambling syndicates, along with the mayor, other leading officials and legislators, each week getting up to two thousand four hundred dollars when the rackets were doing well. What has recently been revealed in New York is mirrored, if often on a smaller scale, in other cities right across the country.

In England the police were launched in a far more favourable atmosphere, when nineteenth century zeal for moral and political reform was at its height. They have had a high reputation for integrity and public service. But they too have had their troubles in the last few decades. For the police of London, as for those of New York, the nineteen-seventies opened with ominous revelations. Two detectives were tape-recorded in the act of trying to extort money from criminals in exchange for favours and by 1973 it was officially admitted that nearly three hundred officers of the Metropolitan police—one in every hundred—had been in serious trouble in the past three years. Now two former heads of departments, together with a group of other officers, who had been concerned in one way or another with the suppression of pornography, have been charged with taking bribes over a lengthy period from purveyors of pornographic material. Nor has London, any more than New York, been alone in this. Elsewhere even Chief Constables have had to be dismissed and some provincial forces have suffered sequences of crime or gross breaches of discipline. All this, like the backcloth of corruption amongst people high up in politics, business and public administration, may have been occurring on a much smaller scale than in the States, but it has had its own impact in so far as it has suggested a decline in standards that were largely taken for granted in the past.

According to some, bribery is so much a part of the traditional way of life in much of the developing world, even in some of the more feudal areas of Europe, that it should not be considered corruption in Western terms. It is a necessary oiling of the wheels, essential to keep things moving. On top of that, over-zealous governments impose numerous, complicated and unworkable regulations, which they expect underpaid police to enforce. Nor is the degree of corruption tolerated in many Asian, African or South American states today any worse than what was customary in Europe at a similar stage of development. Police corruption is only part of the general political and social situation.

It is doubtful whether developing countries would be either flattered or comforted by such attempts to defend them. Corruption at all levels is an acute political issue, on which many governments have fallen. Police venality in the poorer countries may be very understandable but its effects, on individuals, on the national effort, and on respect for the law, can be even more disastrous than in countries where a measure of national unity and material prosperity has been achieved.

Crime and corruption amongst the police, as amongst the rest of us, come in different degrees. The Knapp Commission divided its delinquents into 'meat eaters' and 'grass eaters.' The meat eaters represented the small proportion of policemen who seek huge profits from aggressive misuse of their powers; the grass eaters represent the large number who "simply accept the payoffs that the happen-stances of police work throw in their way."

Everywhere it is the 'meat eaters' whose crimes, when brought to light, hit the headlines. "Wolves patrol the sheepfold" proclaimed the *New York Times*, referring to the admission of a former policeman that he had committed sixteen burglaries with eleven of his fellow police and had stolen regularly from the parking meters he had the task of emptying. A member of the Flying Squad at Nice, caught by another policeman as he was preparing to break into a cigarette shop, cried "Don't shoot me, I'm one of you." It turned out that he was a member of a police gang that had been methodically stealing from householders and shopkeepers calling for help after burglaries or smash and grab raids. The story recalls cases in the United States where police colluded with people who had been burgled in falsifying insurance claims. Even in England there have been occasions when the excellent advice to inform the police when you are going away from home for a time has precipitated rather than prevented a break-in.

But it is the slightly crooked 'grass eaters' who are the heart of the problem. Their sheer numbers make minor corruption appear normal, a part of the 'perks,' even respectable. They build up the code of silence which decrees that exposure is treason. They create a situation in which the new entrant finds participation easier than honesty. And they foster the public belief that the crooked policeman is not the exception but the rule.

Police corruption does not occur in a vacuum. It tends to become pervasive where there is over-extension of the criminal law. It both reflects and responds to corruption at other levels of society. And it works both ways. If the police are corrupt they undermine respect for the law they represent. But corrupt police depend upon a corrupt community. Alongside police corruption there is almost always to be found corruption in government, commerce, industry, labour, the professions and the whole system of criminal justice. There must be bribe givers as well as bribe takers, people who are prepared to pay for their illegal businesses or pleasures, for exemption from prosecution for their motoring offences or other transgressions, people who are prepared to split the fruits of dishonesty or defiance of the law. They may be a comparative handful in key positions; they may dominate sections of cities or whole cities; they may reflect deeply ingrained habits of a whole nation. It is not just that police corruption makes people cynical. People's cynicism makes the police corrupt.

Following the publication of his report, Judge Knapp concluded a public lecture with the words: "This drive has been effective in New York, and on a short range, at least, the police department has responded to it. Now the question is, 'Have we had any permanent effect?' My answer is 'You ask me that fifteen years from tonight'."

Bending and stretching the rules

Almost everywhere the police tend to assume powers not granted them by law, to overstep regulations designed to control them. Underlying this is a conviction that they must have more powers if they are to cope adequately with crime, that the rules work badly, favouring the criminal at the expense of law enforcement.

In England, for example, police are entitled to question people but have no power to detain them short of arresting them for a specific offence. Moreover, far from being able to insist on answers, they must tell a suspect of his right to silence, must allow him, within reason, private interviews with his solicitor, and must not search his premises without either his consent or a warrant. Since in many cases they neither catch an offender red-handed nor can produce willing and reliable witnesses, their best hope is to obtain a confession or find incriminating evidence at his home or workplace. But how are they to get an admission if they cannot hold for questioning, or have to punctuate it with a warning as soon as they see they are getting near the mark? And how are they to get the evidence needed for a search warrant if they believe it is to be found on the premises they want to search?

So they bend the rules. People are taken to police stations to "help with enquiries" under the impression that police have a right to detain them. Cautioning may be delayed or perfunctory. Access to legal advice may be postponed as involving "unreasonable delay or hindrance" to the investigation. Consent to search may be obtained under threat that bail will be opposed or a watch put on the house.

At worst there are episodes like the case of Detective Sergeant Challenor. He was eventually discovered to be suffering from paranoid schizophrenia, but by that time three other police officers stood indicted alongside him with conspiracy to pervert the course of justice and they were sent to prison for it. There had been arrests on inadequate evidence or none at all, outbursts of verbal and physical violence during questioning. And, in one notorious episode, during demonstrations against a visit of the then King and Queen of Greece, Challenor had hauled a young man and two schoolboys off to the police station and planted bricks amongst their possessions with such comments as "There you are me old darling. Carrying an offensive weapon: you can get two years for that."

Challenor may have become a horrible caricature of a bullying and treacherous policeman. The extreme case like this has been comparatively rare in England. But suspicion that police sometimes twist the evidence to get convictions was apparent at the time of the Royal Commission. People who already have criminal records can be particularly vulnerable. Where the police are convinced a man is guilty but know the evidence is thin they may exaggerate to clinch a case.

The role of *agent provocateur* is, in general, regarded with horror in liberal societies. Yet it has been known to crop up in England, much more in the States, especially in attempts to run to earth people who are dealing in drugs. In America, especially, it is often not the police but their addict-informers who are pushed into the odious task of provocation. Yet we must make some distinction between inducing a known criminal to show his hand and exposing a mere suspect to overwhelming temptation. In one case, for example, an English judge refused to hear a detective's evidence against defendants whom he had asked to buy drugs for him, though in another it was held quite legitimate to put marked coins in the pockets of coats left in a cloakroom to entrap a thief known to have been stealing there.

The development of modern techniques has given a fresh dimension to this. Old-fashioned methods of shadowing, eavesdropping, and interception of letters, can now be supplemented by the microscopic bugging device and the tapped telephone. The labourious search through criminal records, the checks with other agencies, gives place to computer checks. In the United States there have been instances of illegal entry of homes or offices by the police without a warrant to instal bugging devices. They have been used in police interrogation rooms and prisons, including rooms set aside for counsel or witnesses. Two-way mirrors, television, filming, have been employed in such public places as restaurants, telephone boxes, cloakrooms, to detect illegal activities of drug addicts or sexual deviants. There have been widespread electronic checks on bookmakers, gamblers and prostitutes on the plea that these could give clues to organised vice rackets. It can be argued that, since modern technology has made the lawbreaker more successful and harder to catch, modern technology must be brought in to check and bring him to justice, that organised crime and political subversion in particular, necessitate such methods. But there are inevitable forebodings where the police use methods which can effectively strike at the very basis of privacy, penetrate the walls of homes, expose the most personal secrets and relationships. It is less that the new methods are morally different from the old kinds of spying, rather that they are far more pervasive and harder to control.

Then there are the new techniques of interrogation. More than half the police departments in the United States use the so-called lie detector in the questioning of suspects. The idea is that lying alters the pulse rate, blood pressure, breathing, in distinctive and measurable ways. A chair may be equipped, without the knowledge of the suspect, to register body-heat or nervous movements, and hidden cameras to photograph changes in the size of the eye pupils reflecting stress. All this can give an illusion of scientific infallibility whilst still leaving ample room for human error in interpretation. If modern techniques could be made more reliable, it might be arguable that they were better than old-fashioned methods of wearing down suspects to extract the truth. Even then, would the automatic conviction of criminals be worth the horror of living in a world where anybody's secrets could be tapped? Nor can we feel very easy about adaptations, however pale, of the range of methods known as "brainwashing." In the more serious, difficult, or potentially dangerous case there is always likely to be deliberate and systematic psychological pressure. Some police text books in the States devote much space to the description of techniques for creating uncertainty and breaking down resistance. Just over half the members of the President's Commission on Law Enforcement, after studying these manuals, concluded that interrogation in the isolated setting of a police station constituted in itself compulsion to confess.

Violence

It is a vital part of the duty of the police to keep violence within bounds, whether the violence of individuals or that of crowds. To do so they must, at times, resort to force. But the necessity, direction, timing and degree of force are matters of discretion. Used for improper purposes, unnecessarily or prematurely, wrongly directed or excessive in nature or intensity, force is rightly classed as violence on the part of the police themselves. Many different circumstances can provoke it. It has been used in attempts to extort confessions, to discover confederates or plans, to prevent further crimes; in a spirit of disgust or contempt towards particular offenders; by way of revenge or punishment, especially when a policeman has been injured or killed; in situations of tension and danger when a crowd seems to be getting beyond control.

England still has probably as little police violence as any country of her size. Her police forces have traditionally been schooled in moderation, both in the control of demonstrations and the handling of suspects. Public toleration of abuses in either situation remains fairly low. Cases like that of Challenor, who struck and shouted at suspects, or of the Sheffield crime squad who flogged known criminals with rhino whips to make them confess to burglaries, or of the senior police officers in Leeds who persecuted a black vagrant, have all rated the headlines for weeks and been cited as atrocities for years.

In the United States the heyday of the 'third degree' coincided with the heyday of gangster violence in the nineteen-twenties and thirties. The use of mental or physical pain to extract confession was widespread. Physical brutality regularly ranged from beating to harsher forms of torture. There was protracted questioning by relays of police under bright, eye-straining lights, deprivation of food or sleep, prisoners were kept standing for hours, and threatened with injury by the mob. This abuse, however, reflected much more than personal callousness or sadism on the part of the police. Behind it was the provocation of feeling helpless against violent and unscrupulous criminals, often protected from prosecution or conviction by corrupt politicians or the terror of witnesses afraid to give evidence against them. To that must be added the problems of a vast country, in which mobile criminals could easily escape, the lack of unified records and action. Violence and brutality by state and local police reflected their weakness, their sense of impotence. The F.B.I., better equipped, better led, better trained, better linked, were least implicated in abuses.

Whereas police brutality to suspects derives much of its menace from isolation and secrecy, police violence against crowds is a public manifestation, whether seen as a warning to rioters or as a ground for complaint. A few countries, notably England, have gone out of their way to make police in charge of crowd control as little like an army as possible, closely limited in possession and use of weapons. Others frankly rely on special riot police, armed and armoured for battle. To challenge these is frankly to risk injury, if not worse. Again it is a circular situation: the degree of violence, potential and actual, rises on both sides at once. There is deliberate provocation by demonstrators as well as excessive or misdirected use of force by police. Police baiting becomes an ideological game for a few protestors, bashing the long-haired an ideological game for a few police. Once that happens in a crowd the excitement, the violence and the brutality can soon infect others. Yet, at least in England and the States, it is ordinary criminals rather than demonstrators who are most likely to injure or kill police trying to arrest or control them, and it is in the course of ordinary investigations that most instances of police violence occur.

The shooting of suspects or demonstrators by the police represents an extreme of violence, an arbitrary dealing out of the penalties of death or mutilation which may strike down the innocent as well as those presumed to be guilty. A girl who had borrowed her father's car without asking permission was shot dead by the police in Germany as she ran away from a road check point intended to catch a bank robber. There have been cases in the United States of police shooting young boys suspected of pilfering because they darted off when questioned and ignored calls to stop. There was likewise the Black Panthers incident, where police called to search a flat in the small hours, responded to a single shot from within by a hail of bullets and killed some of those inside. Or the Attica Prison riot, where ten guards and thirty-three prisoners were killed although no one had been killed by the rioters. In England two Pakistani youths, guileless enough to bring out their toy pistols and make their protest in an embassy protected by some of the few armed members of the Metropolitan police, were shot dead on the spot. Scotland Yard affirmed afterwards that sufficient accuracy to wound without killing could never be guaranteed in such circumstances. At least that caused an uproar. In contrast, it was reported as a matter of routine that over a hundred people, all but three of them black, were shot dead by the South African police in 1972 alone.

In the United States police are armed as a matter of routine, whether on or off duty. An officer who routed an armed robber with a milk crate whilst off duty was commended for his bravery, but almost in the same breath had his pay docked for being without his gun at the time. There are many who argue that the constant carrying of firearms by the police makes them likelier both to kill and to be killed. In the years 1969 and 1970 the Chicago police killed no fewer than eighty-five people, whilst in the twenty years from 1946 to 1966, a thousand police were killed in the United States as against only nineteen in the United Kingdom.

In England the police have been, from the start, regarded as an unarmed civilian body. With the increasing use of firearms by criminals in recent years, there have been pressures towards arming them. This began to become more frequent with the upsurge of violent robberies in the nineteen-sixties. In 1975 guns, real and imitation, were used as many as eleven hundred crimes in London alone. Such crimes as bombings, hi-jacking, kidnapping, together with the fear of attacks upon foreign diplomats, have accelerated the trend in the present decade. Some areas, particularly cities, have established armed units to be called upon in emergencies. The Metropolitan police have their specially trained marksmen. But there is still little enthusiasm for these jobs. The police as a whole continue to resist the idea of being armed in the course of their ordinary duties. The general rule is that guns are issued only for special tasks like guarding people in need of protection from armed attacks, or dealing with particularly dangerous armed criminals. The police view remains that routine arming would lead offenders to use guns more frequently, and would also drive a wedge between police and the public on whose co-operation they have traditionally relied.

Conversely there has been mounting pressure in the United States for tighter restriction on the use of arms by the police. I have even heard a suggestion that they might cease to carry guns regularly in a few areas, if only by way of experiment. But the situation in the States is very different from that in England. The level of violence, above all the rate of murder by shooting, has always been very high and is rapidly rising. The carrying of firearms is commonplace not only amongst criminals but amongst the population at large. As Chateaubriand once observed of the proposed abolition of capital punishment in France, "Que messieurs les assassins commencent"—let the murderers make a beginning.

In England, where the reverse tradition has taken root, there must be the greatest caution in arming the police. We must continue to limit the circumstances in which guns can be carried, define those in which they may be used, do all we can to restrict what is necessarily an evil, even if an inevitable one. You can say of the unarmed English police, as of the English lay magistracy and many other institutions, that they owe their viability and survival to centuries of struggle and painfully built consensus. They will not transplant, but they should not lightly be swept away.

Levels of expectation

It may seem strange that virtually all this discussion of police abuses has been confined to England and the United States, countries which, for all their faults, still have a strong liberal conscience and regard for individual and political freedom. It is such states that are likeliest to investigate police abuses, publish the findings, try to find remedies. Elsewhere they are too often taken for granted, even used as instruments of policy.

The contrast between the Anglo-Saxon and the Continental approach, even in democracies, to this business of investigating public feeling about the police and sifting allegations of abuse is very marked. The modern police of London date from 1829. Six or seven enquiries were held, and their results published, within the first six years of their existence. The modern police of Paris owe their origin to an edict of Louis XIV in 1667. I do not know of any independent enquiry into their conduct since then. Still more marked, of course, is the secrecy surrounding police activity in totalitarian states, where virtually nothing short of the collapse of the regime can drag their behaviour into the daylight, and where to go round asking the man in the street what he thinks of the police would appear, at best, as a bad joke, at worst as a deliberate trap. During a discussion with senior officers at the Ministry of Justice and police chiefs in one South American country I asked how much trouble they had over police beating people. The head of police assured me that no such thing ever occurred. I could see, as I looked round the table, the scarcely concealed smiles on the faces of his own people at this negation of the most obvious realities. When I interrupted with the remark that such incidents could occur even in the most democratic of countries, his answer was "In those places they might—but not here."

In developing countries governments have fallen because of their failures to control corruption, their opponents coming to power on the promise to sweep it away. Its impact upon the poor, who cannot afford to buy justice and may be ruined by extortions, can be more severe than anything suffered in more prosperous areas. Government attempts at social or political reform are defeated. And it is no good pretending that there is no public resentment. One of the most sensitive and sympathetic observers of the struggles of developing nations complained to a chief police officer in New Delhi, with whom he had become friendly, about the local taxi drivers who broke all the traffic regulations. Why did he not order his policemen to enforce the rules? "How could I?" was the reply, "if one of them went up to the taxi driver he might be told 'Get away or I will tell the people that you have asked me for ten rupees.' If the policeman objected that he had not done it, the taxi driver could come back with 'Who would believe you?' "[17]

In the matter of violence towards suspects and trouble-makers, on the other hand, police in less developed parts of the world may be acting in harmony with local feeling and usage. Where it is customary to beat a suspected thief on the spot the police are unlikely to handle him gently.

Why Depend on Professional Police?

Need we have professional police forces at all? Were we not better off when these tasks were the responsibility of the whole community? What about the old English system of constables and watchmen, each serving a year's turn of service to their localities as a matter of public duty? What about the old American frontier tradition of homespun laws, homespun law-enforcers and, if necessary, homespun rope? What about the powers of arrest, investigation and judgment vested in tribal elders or village headmen in less developed parts of the world? What about the peoples' police of the Soviet Union or China? Do not all or any of these keep responsibility for law and order firmly where it belongs, in the heart of the community, in the banding together of citizens for mutual protection and for the reclamation, correction or punishment of wrongdoers?

The theory has always been attractive, and modern attempts to revive such practices have not been lacking even in the West. In parts of the States, indeed, they have never been wholly relinquished, carrying on alongside the regular police, flaring up in response to atrocities, threats, and crime waves, real or imaginary. Sometimes their sponsors take the line of supporting the police, pictured as under-staffed and hampered by red tape in face of threats beyond their power to combat unaided. Sometimes, like the black militants following the Watts riots, they adopt the role of a counter-police, watching the rights of minorities, ready to protect them against brutality or corruption on the part of the regulars. Sometimes, like the Ku Klux Klan, they enforce their own dark laws against blacks, Jews or other hated minorities. Sometimes, as in Carolina and in Brooklyn during the last decade, they have set themselves up to defend such minorities against the Klan.

Amateur policing has three fatal flaws: it tends to prove very dull, it tends to extravagance and it tends to injustice. The 'voluntary' constable of old England dodged his duties because they were dreary and unrewarding as well as occasionally dangerous. The routine of police work and watchfulness is always like that. The enthusiasm of bands of volunteers seldom survives many weeks of uneventful nights. On the other hand the voluntary policeman suddenly confronted with the need for urgent action suffers all the disadvantages of his professional counterpart with none of the safeguards. He is less carefully selected or supervised, far less sure of his powers, less protected by the law, less controlled by training or experience. The chances that he will ignore people's rights or use unnecessary force are higher. So are the dangers that prejudice will overcome judgment, that 'police' functions will overflow into rough justice or, more likely, injustice. Even the deployment of special constables is open to weighty objections.

A 'people's' police—the *druzhiny*—was one of Krhushchev's bright ideas in Russia. There are now estimated to be something like five and a half million of them. They make it their duty to keep an eye on their fellow citizens at work and at play, especially in relation to drunkenness or vandalism. Sometimes they have extended their interest to cases of petty theft, assault or even murder. And a point has been reached where they have themselves had to be restrained from rowdyism and attacks on innocent passers-by. In China the enthusiasm for mutual policing is even more pervasive. There seems scarcely an institution, social or educational, agricultural, industrial or commercial, that does not contain its elements of police. That may produce great benefits in compliance: it leaves scarcely a space anywhere for individual liberty, dissent or even argument.

Nor are the simpler villages or tribal societies of Africa particularly gentle in their own traditional ways of handling suspects. In one such area, asked whether they would beat a thief or hand him over to the police, nearly half the people questioned replied that they would do both. And an observer in the Ivory Coast thus describes the process: "Every bystander, however unconnected with the actual incident, feels himself personally involved, and this general feeling of empathy with the victim gives the public licence to vent its wrath on the alleged malefactor. People surge around the thief, shout insults, rush up to strike him and hurl garbage at him."

No police force, however professional and efficient, can be everywhere all the time. The spread of political and criminal terrorist tactics—the kidnapping, the hi-jacking, the raid on wages in transit, or banks, or wealthy homes—has brought back, even in countries with organised police forces, a measure of private protection. Just as, in the old bad days, the powerful magnate or the wealthy merchant took it for granted that he must live and travel with a horde of servants and retainers to protect him, so the modern millionaire, banker, industrialist, may find himself obliged to pay for private security guards. This must be accepted as a reflex of organised and political criminality. Even so, private security organisations need to be regulated and their role must be carefully defined, especially in relation to that of the regular police. The public must be enabled to distinguish clearly between the two.

There is a growing tendency for householders to feel that they must be armed to defend their families and property. The United States, South Africa, Rhodesia, Northern Ireland, are countries that come to mind, but they are not the only ones. The desire to be armed is fostered by insecurity and tension, the fear that an attack may come at any moment. Yet it feeds the very tension it seeks to counter. The more guns there are about the greater the risk that they may be stolen or used in panic. There is need for very strict gun control and very selective issue of licenses. Unfortunately the possibility of enforcing such legislation effectively is at its lowest where it is most needed.

A third question is whether civilians should join with the police in maintaining public order. That too can be very dangerous. One police chief, rejecting an offer of help in controlling a demonstration against the Vietnam war, observed that he had "no desire to end up fighting two mobs instead of one."

Genuine policing, with all its shortcomings, is expected to take a broader view than that of the private security guard, the householder acting in self defence or the vigilante committed to a particular cause. It is a task which calls for professionals committed to the national interest, to the protection and control of all, recognisable by all, accessible to all.

Safeguards

Given that we have a society which puts a high value upon personal liberty, what safeguards can be devised against abuse of the powers of police?

Some of the most high-principled of police chiefs have claimed that no one is more anxious or better qualified than they to get rid of the 'rotten apples' in their ranks and to check corruption or violence wherever it occurs. The danger of undermining the authority of the senior ranks of police has been one of the principal objections advanced to outside intervention in matters like these. Major revelations of police misbehaviour in democratic states produce the promise of a clean sweep, a fresh and effective system of controls organised from within.

In New York, following the Knapp Commission, a Corruption Control Manual was issued. It listed in detail situations, premises and duties in which corruption might occur, together with instructions on how senior officers should check their men's contacts, warn owners of cafés or building sites against offering bribes or paying for protection, generally try to forestall abuses. This was backed by 'integrity tests,' such as the deliberate dropping of wallets, the planting of simulated narcotics, and agents disguised as hot-dog sellers, to see whether patrolmen would report their finds honestly, accept bribes or try to sell protection. In London, where members of the Criminal Investigation Department had been particularly implicated in corruption, the Metropolitan Police were reorganised so as to reduce the sharp divisions between the detective and uniformed branches and place them under unified supervision. There is an extremely comprehensive set of written instructions on tightening of regulations requiring detectives to notify their superiors of contacts with informers. And a special branch was formed to investigate serious complaints against the police, especially those alleging corruption.

But is all this enough to secure more than temporary improvement? Two things at least make it seem unlikely. A recurrent and sinister feature of cases of gross corruption or violence by members of the police, whether in England or the United States, has been the involvement of supervisory ranks, the delays in reporting misbehaviour, the resistance to investigation, the protracted attempts to cover it up. There was an interval of two years between the reporting of widespread corruption in the New York police and the setting up of the Knapp Commission. A former Inspector of Constabulary appointed to investigate the nature of enquiries being made into corruption amongst detectives in London resigned because, as he put it, "I was beating my head against a brick wall." In the Challenor case there were incidents that must have been known to numbers of other officers: only one reported any of them to a superior. After Challenor had been found to be insane and three other police had been sent to prison, the station superintendent asserted that the verdict did not imply that anything was wrong with his station. When

the Sheffield officers were fined, following a private prosecution, for beating suspects, their colleagues collected the money
to pay the fines. The head of their squad observed, "These things
go on fairly frequently, don't they? You can't have kid gloves in
detecting crime." Of course there are police chiefs determined to
root out abuses and punish the perpetrators. But there are factors
in the situation of the police that tend strongly towards mutual
support and concealment.

Second, for every impulse to radical reform there are years of
inertia, even laxity. There has had to be an enquiry into corruption amongst police in New York on an average every twenty
years. What is most depressing is the repetitiveness of the findings, there and elsewhere in the States. The struggle to prevent
abuses can never be regarded as finished: as part of the struggle
for freedom it demands perpetual vigilance from outside as well
as from within.

From outside the police, courts and judges have taken a hand,
especially in attempting to regulate the ways police deal with
suspects before they appear before magistrates or are brought to
trial. Compare the protections prescribed in England, the United
States and France.

The Judges Rules in England require a caution, perhaps repeated in certain circumstances, and the suspect is entitled to see
his lawyer in private provided it does not interfere unreasonably
with the investigation. But since there is no formal recognition of a right to detain people for questioning, there is no formal
limit on how long they may be so held. It is only after a suspect is
formally arrested or charged that the police are required to bring
him to court within twenty-four hours (forty-eight at a weekend). Nor does a breach of the rules on the part of the police
mean that a confession will necessarily be ruled out as evidence
by the courts, provided they are satisfied that it was not obtained
by unfair coercion.

The American defendant is in some ways more tightly protected, at least in theory, as a result of a series of fairly recent
Supreme Court decisions. It has been held that very prolonged
questioning in itself amounts to coercion. It has been laid down
that the police must notify their suspect not only of his right to
silence but of his right to counsel, free if he cannot afford to pay
for it himself, and that they must not proceed with questioning

unless the suspect has stated that he is prepared to waive these rights. Equally important, the courts are prepared to refuse to accept his admissions as evidence where these rules have been broken.

Under the French system it is the examining magistrate, the *juge d'instruction*, whose task it is to gather and go into information about the circumstances of the alleged crime. But preliminary questioning by the police is inevitable. As a check on this they are required to record the exact date and times of questioning and to get the record initialled by the suspect (if he refuses to do this the reasons must be given). They cannot hold him for more than twenty-four hours without special authority from the Procureur de la République. The suspect, or his relatives, can demand that he be medically examined after twenty-four hours in police hands. When investigation is taken over by the *juge d'instruction* he can withdraw any statement he has made to the police.

Obviously there are loopholes in all these regulative systems. The police can often evade them, sometimes ignore them. Yet it would be going too far to suggest they are valueless. It is rather like speed limits for motorists: the majority exceed them by a few miles an hour when they feel they can do so safely; a few people exceed them habitually by a great deal more than that; others virtually ignore them in what they consider a real emergency. For all that, the legal speed limit, though regularly stretched and occasionally flouted, has a real influence on the general speed of driving in a country. The belief that legal limits, judges' rules, court rulings, have a real effect in restraining the police is demonstrated by the historic significance attached to such measures as habeas corpus. Also by the heat of the current controversies over whether, in our present situation, they should be strengthened or reduced.

Is it possible to set out minimum procedural safeguards to be preserved or established by legislators and courts? I would suggest the following. First, the power to arrest must be restricted to cases where there are serious grounds for suspicion. Second, the police must have no generalised right to search people or premises, tap telephones, use bugs; such powers must be narrowly controlled and subject to explicit authority in each case. Third, the period for which police can detain for questioning on their own authority must be strictly limited. Beyond that the decision on continued detention should be out of their hands. Fourth, the suspect should have the right to dispute police objections to bail, with legal aid if necessary. Fifth, if he has to be kept in custody it should not be that of the police. Sixth, during questioning he should have the right to keep silence. Seventh, he should be entitled to the presence and advice of a solicitor. Eighth, the police should have the duty to inform him of these rights at the outset. Ninth, there must be strict regulation of the procedure for identification parades. Tenth, the retention, as in the United States, of records of arrest and fingerprints in those cases where the suspect has not eventually been convicted, should be prohibited. Eleventh, police should not be expected, as they are in England, to combine the tasks of finding and bringing to justice those they believe to be guilty with the tasks of deciding whether or not they should be prosecuted and in many cases actually conducting the prosecution. Twelfth, and most important of all, independent civilian review should be available whenever there is dissatisfaction with the handling of complaints against the police.

Of course objections and difficulties can be advanced against every one of these safeguards. There are eminent policemen who are deeply convinced that restrictions on questioning and rules of evidence are exploited by clever criminals and cunning, even crooked lawyers. But fears, a few years ago, that the tightening of regulations in the United States would lead to a dramatic reduction in convictions have not been fulfilled. The cynical say that is because the police have ignored the regulations or carried them out in so perfunctory a way as to nullify them. But I know of no evidence, outside totalitarian regimes where the rights of suspects are negligible, that giving the police a freer hand increases their success in bringing to justice the real villains, the professional and dangerous criminals. On the other hand there has been evidence that, in the States, it is defence lawyers rather than the police who persuade suspects that it is in their interests to plead guilty. What concerns a free country is not whether police powers should be limited but where the limits should realistically be set.

Given that there must be rules, runs a second line of objection, they can be effectively devised only by the police themselves. It is unrealistic for legislators and lawyers, academics and judges, remote, calm and detached, to lay down rules the police should observe in the messy, risky, uncertain situations they encounter face to face, calling for quick decisions, putting them in inescapable dilemmas. Those are good arguments for more understanding, more analysis, of the nature of police decisions and the situations in which they take them. I am all for that. But they are not arguments for leaving the police to decide their own limits, settle complaints against their own members. The repercussions of what they do go far beyond their immediate objectives of catching criminals, maintaining order. Those responsible for the substance of law and the administration of justice must take this wider view, exercise this wider control and watchfulness.

However the police are now no longer alone in contending that existing rules relating to police questioning and to evidence tilt the balance too far in favour of the defendant. In England we have had two Lord Chief Justices and the Criminal Law Revision Committee proposing abolition of the right of silence and the caution which draws attention to it. These go back, it is argued, to the days when the accused was undefended, often in danger of death or transportation if convicted. They are no longer necessary nowadays when legal aid is freely available, sentences mild and adapted to individual circumstances. Agreed, people must still be protected against the modern equivalent of rack and thumbscrew—beatings, electric shocks, threats, severe psychological coercion—but that is a different matter. To encourage a man to confess is very far from forcing him to do so. Where information is flowing why risk checking it?

I can see the force of all this but it still fails to convince me. It is a fundamental principle of law in the United Kingdom, the Commonwealth, the United States, that it is for the prosecution to establish guilt, if necessary by independent evidence. It must not rely on getting confessions from the accused. If a suspect is obliged to answer questions put to him by the police, at the risk of adverse comments at his trial if he does not, this safeguard disappears. So long as it remains the need for the caution remains: it is manifestly unjust that the unsophisticated, the suspect who is bewildered and unaware of his rights, perhaps almost in panic, should be left to blurt out statements, true, false or merely confused, that may increase suspicion against him, even result in a wrongful conviction. It has been known to happen, even in murder cases. To take away the right of silence, the stipulated cautions, can do little to increase the chances of convicting the sophisticated criminal, the man who knows the ropes, who has his answers, and even his alibi, well planned in advance.

Quite a different crop of proposals have come from those who believe that the police, even given the right of silence and the obligation to caution, have too free a hand in the questioning of suspects. Some pin their faith on magistrates: the police should be obliged to take a suspect before a magistrate immediately they arrest him; or the Anglo-Saxon system should be replaced by one similar to the French. Unfortunately neither of these meet the problem of controlling police behaviour during the inevitable preliminary questioning. The same applies to suggestions that public prosecutors, rather than the police, should decide whether or not the evidence and the circumstances justify prosecution. That has its merits: in dividing discretion it divides power and the temptation to abuse it. But it still fails to cover the period of investigation, the search for the evidence or admissions to be put before the prosecutor.

The real key to police abuse of their powers in questioning lies in the secrecy of this preliminary period, inadequately defined, inadequately controlled. Not surprisingly there have been various schemes for breaking into this isolation. In some European countries people being questioned by the police are allowed to call in a friend or a relative. In Milwaukee, press reporters can attend interrogations. Or what about a neutral observer, a kind of community representive? It is claimed that, in Milwaukee, not only does the presence of an outsider in no way impede questioning, but the press are less likely to believe allegations that suspects have come under undue pressure. But you could still argue that anyone acceptable to the police who sat in regularly on their proceedings, would be likely to come to accept their standards. Then why not tape-record interviews and have a complete record available in case of dispute or complaint? Unfortunately it is hard to see that this could be a practical proposition in the ordinary run of cases. It may be that, after all, the right to private consultation with a lawyer is the best outside safeguard we can devise at this stage of investigation.

The United States has gone much further than other western countries both in developing modern technical devices for obtaining, storing and disseminating information on individuals and in beginning to seek ways of bringing them under control. With the approval of the Supreme Court there is now statutory control of the use of wire tapping or bugging by the police. They must first satisfy a federal judge that they need these powers to investigate a specified serious crime, that they have reason to believe that messages relating to it are passing through the channel they propose to tap and that other methods of investigation have either failed or proved too dangerous. Similar principles might well be used to restrict access to records, computerised or not, of other social agencies, for instance records of health, employment, taxation, welfare, or of private bodies such as banks.

The so-called 'lie detectors,' 'truth drugs' and the like, I would exclude altogether. In their present forms they are highly unreliable. At most they give access to emotional reactions, to what a suspect finds disturbing, a far wider field than what he may want to conceal as criminal. They are very far from offering objective evidence that he is lying, or combined with the fear that refusal to accept them may be interpreted as indicating guilty secrets, getting him to tell the truth. The mere threat of using them can be employed as an additional pressure on suspects to confess.

Dealing with Complaints

The best rules will be evaded or broken. The best police will sometimes face complaints from the public. The trouble is that, almost everywhere, it is to the police themselves that complaints have to be made. And unless there is convincing evidence of a crime, or the complainant has the nerve and the means to risk action against them in a civil court, the police are final judges in their own cause.

Repeatedly attempts to establish systems of independent review have foundered on the rock of police opposition.

In 1958 the Mayor of Philadelphia launched a Police Review Board, subsequently renamed the Police Advisory Board, to investigate complaints from black minorities in the city. It started with five civilian members, including a Nobel Prize winner and a criminologist. To meet criticism that it did not understand police procedure it co-opted two retired police officers. The police side of the case was put to it by a black police community relations officer. It stressed flexibility and informality in the hearing of complaints, the desire to understand both sides and to promote mutual understanding. In some cases it found a solution in the explanation of police procedure to the complainant, or in an apology for a minor offence. Its maxim was that it was there to help the citizen, not to prosecute the police. Yet the Board is now defunct. Although only a tiny proportion of its decisions were unfavourable to the police, their hostility to it remained unshaken.

The New York Civilian Review Board had a still briefer though not an inglorious life of only four months in 1966. Even that was enough to lead *The Economist* to advocate something similar in England. Its seven members included both civilians and police; one was black, another Puerto Rican. Conciliation and clarification were, again, important aspects of its work. The police and civilian members of the Board came to understand each other's ways of looking at problems. It was argued that, far from diminishing the powers of the police commissioner, the board would strengthen his hand by revealing where grievances arose. In any case the board was restricted to making recommendations: disciplinary action remained within the discretion of the commissioner. Nevertheless a massive publicity campaign by the police, combined with the fear of rising "crime in the streets," produced an overwhelming vote for the board's abolition.

Characteristically, both these attempts to bring civilian adjudicators and conciliators into disputes between the police and the citizen were linked in the United States with the wrongs of racial minorities. Correspondingly the vehemence that insisted on their disbandment just as they were proving their value, had racial undertones. But the fundamental issues are not tied to race, or even to the rights of minorities. Police abuses are the concern of everyone, not only because any individual may suddenly find himself a victim but because they are a misuse of power that cuts at the roots of legality and liberty. In a free country it is the citizen who concedes police powers, the citizen who is entitled, in the last analysis, to demand an account of their use, to judge whether they have been abused. None of the other safeguards proposed can pretend to be proof on its own against misjudgment, evasion, outright defiance. At present the weakest link in the chain of protection is the investigation of complaints.

It is not enough to say that a policeman suspected of crime is dealt with by the courts, that officers from an outside police force are called in to investigate serious charges, that police chiefs are more concerned than anyone to weed out corruption and abuse. Examples of concealment of misbehaviour stick in peoples' minds. It may happen because senior officers have themselves been implicated; or from a false sense of loyalty and solidarity; or even, ironically, from the conviction that the public would lose confidence if police scandals were openly investigated and dealt with. The worst threat to public confidence is the suspicion that they seldom are. It is no good exhorting people to trust the police when, on this score, they frequently do not. Here lies one of the most sensitive points at which justice must be seen to be done.

Police and public authorities have tended in the past to underestimate the weight of public feeling on this issue. The majority of the British Royal Commission brushed aside the idea as likely to weaken the police in their fight against crime. The more recent President's Commission on Law Enforcement in the States, even whilst conceding that wider powers of review might be needed in certain areas, did not recommend any extension of review boards. Yet an investigation in Denver revealed, for example, that nine out of ten policemen were convinced that people did not want a review board, whereas the replies of ordinary citizens showed half the whites and three-quarters of the blacks to be in favour of one. In New York, the Knapp investigators noted that, once they got going as an independent body, many people came to them with complaints on which they had previously kept silent, because they had despaired of a fair hearing. On their recommendation a special prosecutor was appointed to deal with corruption. He had a permanent staff of independent investigators and was in a position both to respond to public complaints and to initiate his own enquiries. As this book went to press we learned that he had just felt compelled to resign.

There have recently been signs that the need for independent intervention at some stage is being recognised by governments, even by some of the police. But the questions remain of the point at which this should come into play, how widely it should operate.

There are strong arguments for independent review of complaints from the outset in areas where suspicion and misunderstanding between police and public are rife. That was the line taken in New York and Philadelphia. Apart from police objections, however, there is the problem of the sheer numbers of complainants to be dealt with. If, in the attempt to counter this, less serious complaints are excluded, one of its major advantages of independent review from the outset, the scope for improving mutual relations by clearing up misapprehensions on both sides, is thereby reduced.

What most people mean by independent civilian investigation, however, is some kind of appeal to an independent body in cases where the complainant is dissatisfied with the way the police have dealt with his grievances. In England a Police Act has just been passed, the first of its kind in any major country, establishing an independent commission to review complaints, other than those involving criminal charges, in which police chiefs decide not to institute disciplinary proceedings. The Commissioner looking into the case would have power to require such proceedings to be taken and, exceptionally, two members of the commission might hear the complaint themselves. This scheme has been designed to provide a double check on disciplinary decisions whilst still leaving the great bulk of complaints to be dealt with by a senior police officer. The debates about it, often bitter, have once again illustrated the fundamental clash between traditionalists who feel that such measures go too far and radicals who feel that they do not go far enough. It would not have passed the wit of man to devise a compromise which could have carried the support of the police yet allayed the anxieties of the public. Unhappily that has not been achieved.

There is, however, a case for saying that it is unfair to single out the police as special subjects of public vigilance or complaints. They are by no means the only officials whose abuse of power may oppress the individual. Those in control of health, housing, welfare or other public services may commit equally grave wrongs. From this point of view the Scandinavian system of an ombudsman, with powers to investigate unmet complaints against all local officials, police included, is far more reasonable. Officials, including the police, can be seen as embodying social conscience, law and order, to the ordinary citizen. But they, in turn, need somebody to play the part of conscience for them.

Not that an ombudsman or civilian review board can hope to do all that is needed. We have repeatedly come across mis-understandings, conflicts of outlook, role and purpose between police and lawyers, police and courts, police and local communities or racial and political groups. These cannot be cleared out of the way by decisions on individual cases. Much more systematic and detached analysis is needed if the police and those who try to regulate or pass judgments on their behaviour are to understand each other any better.

Police research units exist, but too often confine their attention to the technical aspects of police work. Both they and those responsible for police training need to raise their sights, bring in outsiders, try to look at police powers, discretion and decisions in the light of the police role in the total society. That means taking account of the limits on the police and the reasons for them, as well as of their efficiency in the narrower sense.

For more fundamental criticisms and proposals it is necessary to look to independent enquiries, like those carried out by Bar Associations, universities, social scientists. These too can have the faults of their virtues: because they are less inhibited, more detached, they may be more destructive, less practicable. But they reveal, often dramatically, the gaps in the protection of the citizen, whether as a victim of crime or a suspect.

It is tempting to think of abuses of police power and discretion as potential threats to liberty. But it is nearer the truth, and brings it closer home, to remember that the process is by no means one-way. Police abuses are always part of a larger pattern, whether of corruption, of bending the rules, of technology run amok, of violence by individuals, groups or the state itself. The police acquire too much power only when freedom is already in jeopardy, when it has been snatched away or where freedom never was.

CHAPTER

8

Sentencing

What Can the Judge Have Been Thinking of?

Sometimes it is clear to everyone that a judge, in imposing a sentence, has been guided by a single clear criterion. Often, however, it is by no means so clear. The criteria seems to be complex, the reasoning behind them, the choice between them, blurred and hard to discern. Yet the multiplicity of criteria, and the weight given to each, are crucial to the whole business of sentencing.

One August night in 1963 the Glasgow–London mail train was held up, its driver was hit on the head and permanently injured, two and a half million pounds in used notes were hoisted on to a waiting van and taken to an isolated farmhouse for temporary concealment and distribution amongst the members of the gang. The operation had been meticulously planned. Its perpetrators were mature and determined men. On conviction the ringleaders were sentenced to thirty years imprisonment, twice as long as what had up to then been regarded as normal in England for the worst cases of robbery. The crime was described as "an act of warfare against the community, touching new depths of lawlessness." "Severely deterrent sentences were necessary, not only to protect the community against these men for a very long time but also to demonstrate as clearly as possible to others tempted to follow them into lawlessness on this vast scale that if they are brought to trial and convicted commensurate punishment will follow." Both the crime and the punishment were exceptional, dramatic, extreme: they set a new high-water mark for organised crime and its repression in modern Britain. The case of the "great train robbery" has already assumed the character of a legend, almost an archetype.

Yet if the sentences were exceptional the considerations the judges had in mind, both at the trial and on appeal, were not: protection against further depredations by these particular robbers; deterrence of any would-be imitators; retribution proportionate to the enormity of the crime. Long imprisonment could be regarded as the neat response to all three requirements: it would put the miscreants behind bars for a very long time; it would demonstrate that the game was not worth the candle for others; it could be justified as fitting retribution for those who had planned and organised the theft of so enormous a sum and had not scrupled to use violence when a public servant tried to resist them; and it would meet the calculated defiance of law with a calculated weight of legal sanctions.

This is the kind of sentencing most people understand. They want to be protected, they want the law to reassert its authority, and they want the villains to get their deserts. The first two have an obvious utilitarian justification: the need to keep ordinary people and the ordinary dealings of life as safe as possible from those who attack them by force or fraud. But what of the third? Is the question of deserts important only as a modifying factor, a demand that, in justice, punishment shall be applied only to the guilty and shall not exceed the degree of guilt? Have we abandoned as untenable or old fashioned emotional desires to avenge or moralistic arguments for retribution?

Sentencing to seven years' imprisonment a self-styled "Hell's Angel" of eighteen the judge said "You behaved like savages with a callous disregard of the feelings of that fourteen-year-old girl who was half the size of any one of you. . . . You kidnapped this girl from the street and within a few moments she had been raped in that cafe. . . . The girl said you were laughing when it happened. . . . What happened there was horrible and vile." No doubt the judge interpreted the callousness shown as evidence that the youth should be kept out of harm's way for a long time and that other Hell's Angels should be deterred. But his overriding tone was one of revulsion, a feeling that the victim of such an outrage must be avenged, the perpetrator denounced. It can be heard again in the comment of another judge who gave three youths five years' imprisonment each for attacking a couple in a car and raping the woman: "The circumstances are so horrible that they scarcely bear description. Young as you are, the nature of these offences makes it totally inappropriate to send you to borstal or a detention centre."

Retribution, giving the criminal his deserts, may have a more civilised ring than revenge, with its passionate refusal to recognise limits. Yet it stirs feelings that run just as deep. Impatient with the "ramblings" of penologists about the failings of modern penalties, a victim of crime wrote to a newspaper: "They have left out the pre-eminent purpose of punishment. Those of us who try to do right need to see something unpleasant happen to those who do wrong. It is not a matter of revenge or deterrence or reform. It is a matter of satisfying the inner longing for justice. I was once burgled. The police efficiently caught the burglar. He pleaded guilty and asked for twenty other burglaries to be taken into account. I sat stunned at Quarter Sessions while the Chairman addressed him in kindly and encouraging words and put him on probation. Nobody spoke to me—I was only the victim. The man who had stolen my peace of mind left the court on his wife's arm and smilingly raised his hat to mine. My respect for the English courts has never recovered from that morning's experience."

More to his satisfaction would have been the pronouncement of an Appeal judge on the representations of a court official, jailed for four years for fraudulent conversion of court funds, that such punishment was not needed as a deterrent since it was most unlikely that others in his position would follow his bad example: "This sentence is not a deterrent sentence, it is a sentence which is fully merited, in the opinion of this court, as punishment for very grave offences and as expressing the revulsion of the public to the whole circumstances of the case."

That rather formidable old nineteenth-century judge, Sir James Fitzjames Stephen, who had considerable experience of sitting in judgment, made no bones about it. And he had in mind not only the horrors of murder and rape but the run-of-the-mill theft. Holding it right and proper to hate criminals, he conceded that the public desire for vengeance could go too far. Yet in his eyes the unqualified condemnation of such feelings was as ill-judged as an unqualified condemnation of the sexual passions. The execution of criminal justice gave legitimate expression to the one, as marriage did to the other. Or, to use another of his similes, "the sentence of the law is to the moral sentiment of the public in relation to any offences what a seal is to hot wax. It converts into a permanent final judgment what might otherwise be a transient sentiment."[18] "You have been found to be dishonest parasites on the financial system of this country. You have also been found to be dangerous parasites because your action strikes at the base of all commercial and financial honesty and all understanding between men." That denunciation was addressed to two businessmen who had perpetrated major frauds in the City of London. "You destroyed the law when it required to be preserved and weakened its authority when it was your duty to support and assist." Two English police officers were sentenced to seven years' imprisonment for accepting bribes and conspiring to pervert the courts of justice, two others for hounding a vagrant. In Turkey a similar sentence was passed upon a writer for translating and publishing the works of Marx and Engels. In Russia the manager of a mechanical repair shop was sentenced to death for theft of state property. In the Philippines a Chinese businessman was condemned to public execution by firing squad for trafficking in drugs. In Nigeria something like eighty people suffered the same fate within a year or two for armed robbery.

All these sentences had, of course, their elements of deterrence and retribution. But they have in common another element, what has been called denunciation, a powerful reassertion or assertion of the values attacked: in England financial integrity, faith in justice and the incorruptibility of those who enforce it; in Turkey the rejection of communism; in Russia the sacredness of communism; in the Philippines a new resolution to reform a 'sick society'; in Nigeria the vindication of law and order after a nightmare of civil war.

Of course both deterrence and retribution come into it as well: it is hard to find an example of 'denunciation' where this is not so. Indeed it has been suggested that the very concept is superfluous. Yet Durkheim believed that precisely here lay the only real justification of punishment: it seldom reformed criminals, it certainly did not eliminate crime, but it did maintain and reinforce the values of a society. This kind of consideration is somewhere in the mind of a judge who passes an 'exemplary' sentence. True, he hopes to deter, to administer a salutary shock which will check racial attacks, or mugging, or pornography for the young, or vandalism in trains or telephone kiosks, or shoplifting, or baby-snatching, or whatever is the current cause of scandal. But he also hopes to reassure and reassert.

Correspondingly there is the situation where a judge is conscious that the values presented by the criminal law have already lost much of their hold, that public opinion has moved so far that denunciation is out of place. A woman doctor in Holland who had yielded to the pleas of her dying mother and given her a lethal dose of morphine, escaped with a suspended prison sentence of one week, the Chairman of the court expressly stating that he had taken into account the fact that many Dutch doctors no longer accepted that life should always be prolonged to the bitter end.

At the other extreme are the cases where severity of sentences rises simply because a value held by the majority is coming under strong attack by a minority, or even in an effort to convert the majority to a new outlook. Both apply, in different parts of the world, to sentences designed to stem the flood of what are usually classified as dangerous drugs. In England and the States these still appear as comparatively new threats. In some parts of the East they represent almost a traditional way of life, now rather suddenly classed as reprehensible. Yet in either situation heavy sentences upon manufacturers and traffickers, often upon users as well, have been used in the attempt to drive home the enormity of the danger. A prison sentence is the normal penalty for trafficking in many European countries, and some have doubled the maximum since 1970, giving maxima of ten or twenty years. In Italy possession, sale or smuggling incurs a minimum of three. Turkey was imposing the death sentence for manufacturing, a minimum of two years for possession and up to thirty years for trafficking, whilst still making some two million pounds per year from the state marketing of opium.

The sentence on Timothy Davey, who was just fourteen when he was caught doing a deal over twenty-six kilos of hashish in Istanbul, caused an uproar because it brought a head-on clash between two contrasting concepts, especially pronounced in dealing with the young. "It seems absurd and inhuman," was the instant reaction of *The Times*, "that any boy of fourteen should be sentenced anywhere to six years plus three months imprisonment." A day or two later another leading article in the same paper looked at it from a different angle. "Turkey and other countries are understandably tired of being plagued by the products of over-permissive and over-indulgent societies. Those who have been quick to label the Turkish court's sentence 'barbaric' might pause to remember the 'barbaric' effects which drug addiction is having on growing numbers of young people."

The maxim that the first concern in dealing with children and adolescents must be their own present and future welfare has been pushed further and further in civilised countries. Under the Turkish system Timothy was to spend most of his sentence in a reform school, not a prison. In England he might merely have been committed to the care of a local authority. Re-education, rehabilitation, an eye on the future rather than the past, have become the watchwords in sentencing the young, when their offences are grave enough to require them to be sentenced at all. They have likewise been the continuing guidelines in carrying out the measures the courts may decide.

Graham Young, also a boy of fourteen at the time of his conviction, had given poison to his father, his sister and a school friend. Psychiatrists warned the court that he was extremely likely to do it again if he got the chance, and was too dangerous to be sent to an ordinary psychiatric hospital. So he was sent to Broadmoor with the stipulation that he should not be released for at least fifteen years except at the express direction of the Home Secretary. As he grew up and the psychiatrist responsible for his treatment became convinced that he was cured, the conflict between detaining him for the protection of others and releasing him so that his rehabilitation could be completed became acute. He was released after nine years, got a job in a chemical laboratory, killed two of his workmates and poisoned others before being brought to court again and sentenced to life imprisonment for murder and attempted murder. The circumstances were, of course, both extreme and rare. The overwhelming majority of murders are never reconvicted after release. Of three hundred people who left Broadmoor in the years 1959 to 1971 only one other committed homicide whilst on parole.

In England a judge imposing a sentence of life imprisonment

for murder or manslaughter can make a recommendation about the minimum number of years the killer should serve before being released on parole. Like a restriction on release from a mental hospital, it is not binding upon the Home Secretary, but it is influential. A fairground worker said to be a psychopath and with many previous convictions, some for sex offences, was sentenced to life imprisonment. The judge said that in his opinion this psychopath was incurable and "grave responsibility" would be incurred by those who released him. But a diagnosis of mental abnormality is not the only circumstance that may convince a judge that an offender is too dangerous ever to be at large again. The professional criminal who cold-bloodedly uses violence as a tool may get similar treatment. "It is quite obvious from your record," said another judge to a young man who, whilst on the run from Pentonville, had shot a detective dead, "that you set yourself at an early age to be a total enemy of society. You are a highly dangerous man who resorts to violence without any compunction or hesitation whatever." He recommended that in this case the sentence of life imprisonment should mean life. In all these sentences the personality of the offender loomed as large as the gravity of his crime. The judges justified exceptionally lengthy confinement not primarily in terms of retribution or deterrence but of dangerousness.

Yet a study of the criminal career and future prospects even of a persistent adult offender may lead a judge to a very different conclusion. Another prisoner, like the burglar whose jaunty exit so outraged his victim's sense of justice, had been put on probation for breaking into someone else's premises. Having subsequently committed two further offences, he was brought back and sentenced to three years' imprisonment. He had the temerity to appeal and, probably as much to his own surprise as anyone's, another three year term of probation was substituted. "If he is not enabled to make use of probation on this occasion," reasoned the court, "he is the sort of man who will spend the rest of his life in prison." The attention of the court in that case had shifted drastically from the gravity of the offence, the prisoner's deserts, the need to deter others from burglary, to consideration that this might be the last chance of rehabilitating him and a calculated gamble on the faith that "the best protection for society is the reformation of the offender."

Is There any Overriding Principle?

Does it make sense that those responsible for sentencing should try to achieve such a variety of ends, follow so many different principles? Faced with the train robbers they were concerned with preventing repetition of the offence, by them or by others, with ensuring that the depredators suffered for their crimes and had no chance to enjoy the fruits of them. Faced with the rapists and the brutal, considerations of retribution, denunciation, even perhaps revenge, predominated. Faced with breach of trust or abuse of public position, they emphasised the vindication of the law, the redressing of the moral balance. Faced with grave offences committed by the young, there was conflict between the claims of deterrence, retribution, prevention and hopes of reclamation. Faced with the recidivist who had come to his 'last chance', there was again the tug-of-war between his past record and his future prospects.

Yet a sentence out of all proportion to the crime is repugnant. After all, it is the offence that puts its perpetrator in the hands of the judge, it is in response to the offence that measures of punishment or treatment are imposed. The sentence must be warranted by the crime. Hence the scales of penalties allowed by law for offences of varying gravity. Capital punishment might eliminate cycling without lights (an offence dangerous to life and limb) but it would be unthinkable. The threat of a small fine might be no more ineffective than that of life imprisonment in preventing some kinds of murder, but that too would be unthinkable.

How you can balance the two very different things, the crime and the punishment, is hard to say, but the feeling that there should be some balance is ineradicable, and it affects sentencers as much as legislators. At the beginning of this century the English judges drew up a scale of what they considered 'normal' penalties for a range of common crimes: not the maxima permitted by law but the sentences that could be considered appropriate in the absence of special circumstances of aggravation or mitigation. In 1970 the Magistrates' Association, recognising that many Benches

already had their own guide lines in sentencing motoring offenders, attempted a scale of 'normal' penalties for offences of
this kind. They denied any desire to establish a tariff of penalties,
but they hoped that, by adopting similar starting points in sentences, to be reduced or increased according to circumstances,
they might achieve greater consistency. These efforts to put into
writing some consensus about the levels of punishment appropriate to different crimes may have been less helpful than their
authors hoped. The perplexities, conflicts and disagreements of
sentencers cannot be so easily solved. But they testify to an apparently ineradicable sense that as wages should be to work,
punishment should be to crime: the more difficult or valuable the
work the higher should be the pay, the worse or more damaging
the crime the severer the punishment. In both fields, whatever
may be argued to the contrary, there is a customary level of
expectation, which changes only gradually. In a court the level is
known to judges, barristers, probation officers, to all but the most
naive offenders. Any sudden departure from it, even in a case like
that of the train robbers, produces a sense of shock. Prisoners
seeking parole will argue that they have done enough time for
their particular crimes, parole authorities may feel that for the
same reason they must do more before they can decently be
released. The idea of 'just deserts' persists right through. A letter I
recently received from a perplexed and conscientious English
high court judge illustrates one part of the dilemma.

"The longer I am a judge the less confidence I have that I
understand what I am trying to do when I am sentencing. I
suppose that all judges assume we are trying to deter from further crime (a) this person, (b) other people. But I am becoming
more and more sceptical of the effectiveness of either. If we
reduce our sentences by excluding the deterrent element, there is
no doubt that we alarm the public and decrease their confidence
in us. Last Term I passed what I thought the right sentence on a
young man of nineteen for robbery and manslaughter (six years)
and had several really vicious letters complaining that I had not
given more. This case was complicated by the fact that the defendant was black. Obviously, one cannot be looking over one's
shoulder at what bigotted hangers and floggers are going to feel
about one's sentences, but I suppose one has to give public opin-

ion some weight, even though I think it is usually wrong on these things."

For two hundred years there have been endeavours in various countries to establish a coherent set of principles. But on the fundamental question of what should take precedence in sentencing—retribution for the offence with its assumed power to deter, or measures of treatment for the offender with their assumed power to reform—we remain as divided as ever. Lawyers and philosophers disagree. Judges and magistrates disagree. People in all walks of life disagree. Two polls held in England in the late nineteen-fifties, one by the B.B.C., one by Gallup Polls, coincide to illustrate this. One in three of those questioned thought the primary purpose of sentencing should be to punish the criminal in order to deter others. One in three (or one in six, according to how the question was put) thought it should simply be to punish him for what he had done—straight retribution for the crime. And one in three thought it should be to reform or reclaim him. Similar differences, though not necessarily in the same proportion, have been found in the attitudes of magistrates and judges. These differences seem to run very deep: no amount of argument is likely to resolve them. They are comparable in this to divisions over the death penalty.

Few people, however, would take up an absolute position under any of these heads. However convinced they may be that the offender should be sentenced on the basis of deterring others or on the basis of retribution for his crime, they would begin to flinch if he was very young, or under great pressure, or not fully responsible. However convinced that reformation should be the overriding objective of sentencing, they would waver when faced with the unrepentant criminal careerist or the abnormal killer whose release could be allowed only at grave risk to others. So the question for the sentencer cannot be "What should be the primary objective in all cases." If he asks that he will get answers like "The prevention of crime" or "The protection of society," which beg the very question he asks—prevention and protection by what means? The real question is, which of these paths should be taken in this particular case? What predominates in one will not predominate in another.

The emphasis in sentencing shifts with the kind of offence.

Retribution and deterrence tend to be the ruling motives at the extremes: the worst offences which will incur substantial terms of imprisonment whatever the age or past record of the offender, and the trivial offences which will incur run-of-the-mill fines. Consideration of the offender takes priority mainly in the middle range, where the compulsions to make an example, to protect, to deter, are less powerful yet the offence is serious enough to justify investigation of the offender's background and treatment adapted to his individual needs. Here the ruling factor is the condition of the offender rather than the gravity of the offence. His age, his criminal record, his attitude, responsiveness, prospects, all go into the scale. Not that this kind of sentencing will always and necessarily mean a 'softer' outcome for the offender. He may get discharge or probation. But it may be decided that the only hope of redirecting him, or of getting him away from bad influences, idleness, drugs, alcohol, is a period in a borstal, training institution, hospital, prison. Or his personality and circumstances may point to an extended sentence for the protection of others.

This brings in another complication of sentencing. Whatever the main objectives of a particular sentence, the others cannot safely be ignored. However convinced the court may be that general deterrence is essential, it cannot afford to lose sight of justice, which is the other face of retribution: an 'exemplary' sentence which goes too far beyond what has been customary for the offence can produce a sharp sense of injustice. However convinced of the need for severe retribution, or for protecting others from an offender, the court cannot wholly lose sight of his chances of eventual rehabilitation. Promises that he will receive treatment or training in prison, whether or not well-founded, bear witness to an underlying consciousness of this in judges and magistrates. Whatever the primacy given to rehabilitation, the court will still have somewhere in mind considerations of deterrence and retribution. Warnings of what will happen if the chance is abused, considerations of whether probation can be allowed to one partner in crime if it is denied to another, the new measures provided in England to require work or attendance for training as a condition of release, are in one sense evidence of the feeling that neither offenders nor the public should be allowed to feel they are simply getting away with it.

There are other limiting factors in sentencing, and indeed in the whole matter of punishment. One is humanity, what society is prepared to accept at a given stage. Whether you consider it the duty of the judge to retaliate for the offence or to protect the society by preventing its repetition, you cannot go the whole hog. 'An eye for an eye' was originally a law designed to limit vengeance, to replace it by just proportion. There was world-wide revulsion against a group of Arabs and Japanese who seized a large passenger plane and threatened to blow it up with every-one aboard. But when Libya announced that they were to be tried under Islamic law, and it was recalled that this might bring sentences of death, or the amputation of a hand or foot, there was another kind of revulsion. The revulsion was illogical, in that mutilation could be seen as a lesser penalty than execution, but it was logical in the feeling that it would be a step back to bar-barism.

Yet another limiting factor is what Jeremy Bentham called 'frugality'. To him punishment, like crime, was an infliction of evil. The criminal law and the judge should impose no more of it than was necessary to achieve the end in view. On an earthier level there is the simple question of economy. Why resort to measures like imprisonment which are expensive in money, build-ings, staff, if measures like fine or probation, which are less costly, are equally likely to deter or reform? The whole level of sentencing in these respects differs between different countries, and different parts of the same country, just as it has differed over successive periods of history, without demonstrably differ-ent results. There is no magical or established degree of severity calculated to achieve either 'just retribution' or effective deter-rence. It is a question of what we have become used to and feel able to accept. The rise in crime has put pressure on everyone, including those responsible for sentencing, to prefer the more 'economical' to the more expensive measure.

Many other principles can be teased out when philosophers and lawyers are let loose for years on end to analyse the business of sentencing. But to the judge sentencing is a practical business, often carried on in circumstances over which he has no influence at all. Such mundane matters as the overloading of the courts and the penal system: the injustice of prolonged detention pending trial: too few probation officers or too few places in specialised institutions: increase in crime, particularly in its dangerous forms, creating social pressure for swift and exemplary justice: any of these can limit the kind of sentence that can realistically be imposed.

Disparities in Sentencing

A sentence may appear misconceived, improper, unfair, for several reasons. It may seem too soft a response to the crime: a recent Gallup Poll in England showed that six out of ten people thought leniency a very important cause of crime and violence, eight out of ten that it was at least fairly important. And the Lord Chancellor has repeatedly lambasted magistrates' courts for lack of resolution in imposing stern penalties where required. Alternatively, a sentence may appear too harsh in relation to the offender: outcries against prison sentences imposed on girls who carried off other women's babies have been evidence of that. Yet again, the sentence may simply seem out of proportion to the offence, like sending to prison a bus conductor of previously good character for swindling the bus company out of a trivial sum.

A concise, though not necessarily exhaustive, summary of the kinds of improper disparities in sentencing that can occur in the United States was given by Whitney North Seymour, a former public prosecutor, at a recent Judicial Conference in New York. He distinguished four levels: the disparities between individual judges; the disparities between offenders convicted of the same offence with essentially the same factors present; the disparities related to kinds of offence, especially the milder penalties meted out for white-collar crimes; the disparities between the state and the federal systems.

One definition of unfairness on which everybody would agree is that the sentence is out of proportion to the offence, out of proportion to sentences meted out to others for similar crimes, and cannot be justified by any special characteristics of the offender which might legitimately have been taken into consideration, such as his age or mental condition. Some differences in sentences are attributable to real differences in the offences, offenders or situations, which often go unmentioned or unexplained in press reports. Not every public outcry against a judge is justified. But some discrepancies stem from underlying causes which are less defensible, even rather sinister.

Uniformity in sentencing is not the answer. If we could achieve it it would still be unfair. It could be justified only if we were faced with identical offences, and identical offenders. What we need is consistency in principles and proportions, means of countering prejudice and bias.

That unjustified discrepancies in sentencing exist there is no doubt. They have been demonstrated again and again, in terms of areas, of courts, of individual judges and magistrates. The North of England, for example, uses probation more sparingly, fines and imprisonment more readily, than the South. Federal courts in Brooklyn impose sentences for robbery averaging half as long again as those imposed in Manhattan. A recent analysis revealed that the proportion of robbers in Birmingham getting more than four years was six times as high as that in Bristol. Magistrates impose far lower fines for motoring offences in some English courts than in others. We cannot immediately jump to the conclusion that the whole of these differences stems from unjustified prejudice, harshness or ignorance on the part of the courts. It may be that, in general, the North of England has to contend with more and tougher criminals than the South, that robberies in Brooklyn are worse than those in Manhattan, that lower motoring fines in rural parts of England may quite properly reflect less serious offences or lower incomes. But even when allowance has been made for such factors, evidence remains of unjustified discrepancies. It can make quite a difference whereabouts in the country you commit your crime.

The existence of unjustified discrepancies is borne out by still more localised comparisons. For example, a survey disclosed that a boy or girl coming before the juvenile court of one northern English town had an eight in ten chance of being put on probation, a boy or girl in a neighbouring town only a one in ten chance. Admissible differences in their offences or circumstances could hardly have been great enough to explain that away. Similarly very considerable differences have been found in the proportion of cases in which various English magistrates' courts resorted to imprisonment. Discrepancies still come to light when investigators get right down to the opinions and practices of individual judges or magistrates, even within a single area or court. "Hanging judges" may be out, but differences in the importance attached to the various objectives of sentencing, in general severity, in the attitude to particular kinds of offence and particular kinds of penalty, persist. Some part of discrepancies stems from deeply rooted differences in the personalities, attitudes and backgrounds of the sentencers, be they judges or lay magistrates.

Most sinister is the charge that disparities in sentencing can be ascribed to racial or class prejudice: that black people in the United States or South Africa, for example, receive severer sentences simply because of their colour; or that people from the slums are more likely to be sent to penal institutions, and sent for longer terms, than more prosperous people who have committed similar crimes. More specifically, some researchers suggest that blacks are more likely to be sentenced to imprisonment than whites convicted of similar crimes, and that a robbery or rape committed by a black against a white incurs a severer punishment than a robbery or rape in which victim and assailant are both of the same colour. Others reply that differences occur because the blacks are likely to have worse criminal records or to have committed the crimes in more aggravated forms. Naturally this issue runs through the whole spectrum of criminal justice. If there is discrimination it starts long before the stage of sentencing, with police decisions to question and arrest and the ability to obtain good defence attorneys. In such matters the dice may already be loaded against the blacks by the time they appear in court. That differences exist cannot be denied. The question is whether they result from judicial prejudices, or from real differences in criminality, legitimately taken into account, or from both.

It is when it comes to death sentences for such crimes as murder and rape, that the charge of racial prejudice is hardest to refute. The Florida Civil Liberties Union found that between the years 1940 and 1964 almost as many whites as blacks had been convicted of rape: one hundred and thirty-two to one hundred and fifty-two. But only six of the whites, as against forty-eight of the blacks, were sentenced to death; only one white, twenty-nine blacks, were actually executed. The most disturbing piece of evidence that there has been discrimination in capital cases in certain parts of the United States is that published by Professor Marvin Wolfgang and Mark Riedel in 1973.

In South Africa Dr. Van Niekeuk of Natal University was charged with contempt of court in imputing improper motives to the judges of the Supreme Court, "namely that they discriminate unfavourably against non-whites in the imposition of the death penalty." He claimed that death sentences for rape and murder in the years between 1947 and 1969 had borne no relation to the numbers of each race convicted of these capital offences. Of a hundred and twenty whites convicted of raping non-white girls and women none was sentenced to death, though three others got such sentences for raping white women. Meanwhile a hundred and twenty black or coloured men were condemned to hang for rape. One judge, writing in Afrikaans, quite frankly explained "for a white woman rape, particularly rape by a non-white, is a terrible experience. For the majority of Bantu women rape, even by a white, is something which can be compensated by the payment of cattle." People who live in American slums sometimes complain of similar discrimination by courts in dealing with assaults in their areas, alleging that such offences incur small penalties since they are regarded as part of the natural way of life of the very poor.

Murder, however, can hardly be brushed aside as a minor crime. In South Africa black and coloured people are seven times as likely as whites to be convicted of murdering someone of different colour. But the proportion executed for such murders is much higher: fifty to one. It was to the credit of the South African court which heard his case that Dr. Van Niekeuk was acquitted, though, on the figures, what other verdict was possible?

Members of a depressed or oppressed section of society may suffer direct discrimination in sentencing because the judge shares local prejudices. But they may suffer them simply as an integral part of their other disadvantages. Because of poverty they may, even under systems designed to give equal justice, have less chance of getting bail for lack of a fixed address or respectable referees, and if they have to depend on free legal aid it may be granted only a short while before their cases are heard. The chances of bringing out mitigating circumstances are thereby reduced. Even without judicial prejudice, the sentence may thus be more severe than that of the well-heeled defendant who can secure counsel early. That is a reminder that the judge is not the only one to influence sentencing.

The discrepancies that most commonly hit the public eye are cases where, for example, a pair of apparently similar offenders, having committed apparently similar crimes, receive widely different sentences around the same time. A former Director of the Federal Bureau of Prisons in the United States tells of two men sent to prison for fraudulently cashing cheques. One had obtained about sixty dollars: it was his first offence, he was out of work, his wife was sick and he needed the money for essentials. The second had obtained thirty-five dollars: he had been in trouble before for offences of drunkenness and failure to support his family. There seemed no significant differences for sentencing purposes. But they had appeared before different judges: the first got fifteen years, the second thirty days. In Britain, two girls of similar age, both first offenders and both admitting knife attacks, appeared at two different courts. One, shortly after the last war, had stabbed her boy friend when she saw him with another woman. She was put on probation. The other, much more recently, had stabbed another girl who had just insulted her in a cinema. She was sent to borstal, the judge remarking that it was nonsense for a psychiatrist and a probation officer to suggest that she be allowed probation for such a crime.

What makes such differences possible? First there is the latitude allowed by the criminal law. It sets a maximum penalty for a particular class of crime, but that maximum tends to be pitched to meet the worst conceivable instance: the most deliberate embezzlement involving the most despicable betrayal of trust; the most savage and unprovoked assault producing the severest and most permanent injuries short of death. At the same time there may be a general provision, as there is in England, that a court may use discharge, probation, suspended sentence, in any case where the sentence was not fixed by law (virtually for any offence short of murder). Between the extremes of a maximum penalty that may be as high as life, or perhaps thirty years, in prison, and a minimum of discharge, the discretion of the judge is obviously enormous. And much depends on the way he interprets the significance of the legal maximum. It may well be that one of the two judges sentencing the two American embezzlers saw it as a kind of penal atomic bomb, an ultimate weapon to be used as a last resort, whilst the other saw it as a standard means of defence against this particular crime and had no hesitation in imposing it for a run-of-the-mill case.

In countries like the United States, with a federal constitution under which each state has its own criminal laws, its own penal code, there is an additional factor: wide differences between the legal maxima prescribed and the latitude allowed in different states. The maximum penalty for the same offence may be a fine or a year in prison in one state, fifty in another. Though smaller countries, with unified legal systems, may not have to contend with that particular kind of anomaly, they suffer from others. The criminal laws of England were not all made at the same time. Ideas on the relative gravity of different offences, the severity of the maximum penalty required, have varied over the years, and will no doubt continue to vary. So the upper limit of judicial discretion may often be either higher or lower than our modern sense of justice finds acceptable. This last can apply even in countries whose penal codes were devised and promulgated as a whole, in the attempt to produce a coherent scale of penalties: changing social conditions, changing moral attitudes, new offences, new penalties, can still upset the legal balance.

The decisions of judges and magistrates in sentencing are not governed only by the maximum sentence available to them or the range of alternatives from which they can choose. Given the latitude allowed them by law, they may also have very different ideas about which of the possible objectives of sentencing should be given priority in a particular case. In dealing with this man, this girl, should they concentrate mainly on general considerations of retribution warning off others, or should they concentrate mainly on the individual, whether in the hope of reforming him or in the anxiety to keep him under restraint for as long as possible? It was clearly a deep divergence of this kind that produced the very different decisions about the two girl stabbers: the first sentencer concentrated upon the chance of a new start for the offender, the second was convinced that violence must be severely punished, whatever the consequences to the girl's future. An element of unpredictability, of unfairness as between individuals, has been part of the price of broadening discretion, of encouraging courts to look at and consider the offender as well as the offence. Inevitably there will be questions as to whether they have put the emphasis on the right place in a particular instance, as to why they chose one course for one defendant, a very different one for another.

So we reach the point of recognising that, if a court found us guilty of, say, shoplifting, we could be very uncertain of what sentence we should receive: perhaps a fine; perhaps discharge or probation; possibly prison, perhaps for a few days, perhaps running into months; perhaps some quite new-fangled order, like community service. If our crime was grave, or our record bad, the extremes would be even wider: we might feel it was a toss-up between probation or many years in prison. This kind of diversity and uncertainty has been made possible precisely by what have been considered major achievements in penal reform over the past century: more discretion to the courts; a wider range of criteria in sentencing; above all the importance of taking into account the offender as well as the offence and extending the choice of measures available for dealing with him.

Limits of Discretion in Sentencing

"All the judge should have to do," claimed the author of the first criminal code thrown up by the French Revolution in 1791, "is to open up the law and find there the punishment applicable to the crime that has been proved; his duty is simply to impose that penalty."

This refreshing simplicity was the culmination of the dream of eighteenth-century European reformers, like the Italian Beccaria, the Frenchman Montesquieu. Their revulsion against the chaotic, cruel and arbitrary punishments of the despotic regimes under which they had lived convinced them that the keywords of sentencing should be certainty, moderation, proportion. The revolutionaries took them literally. But like most extreme positions, this proved untenable in practice. It is impossible to foresee and define, down to the last detail, all the possible forms an offence may take, all the imaginable degrees of its gravity. And even if that could be done it would still be impossible to devise or maintain a right scale of penalties that would, at the same time, be just and have equal impact upon all offenders, regardless of their age, circumstances, personalities. Inevitably anomalies appear, legal ingenuity combines with humanity in successful attempts to get round them, and the very attempt to achieve absolute certainty in sentencing results in new kinds of uncertainty. In short, a system that tries to eliminate judicial discretion is neither just nor certain, even in terms of the offence, let alone of the offender. Such a system must depend for its justification, if justification it has, solely upon hopes of general deterrence, on the assumption that, if people know the exact price of a crime, they will not risk having to pay it.

Even the most rigidly legalistic systems of criminal justice on the Continent of Europe have found it necessary to accept such concepts as mitigating circumstances and the suspended sentence, and the tendency has been for the escape hatches to be enlarged. Indeed, just at the time when voices are being raised in Scandinavia and the States to demand a retreat, in the interests of justice, from excessive emphasis on treating the offender rather than the offence, from undue infringements of liberty in the name of rehabilitation, in countries like France and Belgium there have been moves in the opposite direction, towards more concern with the needs of the offender, less with the meting out of justice to fit the offence alone.

Extreme views, on one side or the other, may be as necessary to advance as lifting up first the left foot and then the right is necessary to walking. But they will not get us far in practice unless we keep a balance. A whole range of devices for the statutory control of sentences are in use in different parts of the world. Which of them encourage coherent patterns without at the same time imposing too much rigidity, control discretion without destroying it?

First, the options. There may be mandatory penalties, where the law gives the court no choice: if, for example, the prisoner is convicted of murder he must be sentenced to life imprisonment. There may be minimum sentences, where the law requires the court to impose, say, at least a year's imprisonment, or disqualification from driving, but leaves it to the court's discretion whether to impose more but not less. There may be minimum sentences in another sense, commonly used in the United States though not in England, to regulate restrictions on parole. In some countries, and for some offences, the period to be served before parole can be considered is at the discretion of the judge, his 'minimum sentence'. In others the proportion to be served is laid down by law. In England, for example, it must be a third of the sentence or a year, whichever is longer: the judge has normally no say in this. The most common of all restrictions on sentencing, however, is the maximum penalty: for such and such an offence the court can impose no more than such and such a penalty, though it may, at its discretion, impose much less.

There are other devices that may be used to circumscribe the sentencers, especially in their choice between imprisonment and freedom. If you commit certain kinds of offence, have already been convicted a certain number of times, are above a certain age, you may be excluded from fine, from discharge, from probation or suspended sentence: the court may be allowed no option but to send you to prison. Or it may work the other way round. There are minor offences for which you cannot be sent to prison, at least unless you persist in committing them over and over again. There may be restrictions related to age or previous good record. There are limits on the length of sentence, and the level of fines, that can be imposed by summary courts. There may be conditions as to the age, mental condition, criminal record, current offence, that permit a court to sentence a persistent or dangerous offender to an extended term of confinement.

All these legal conditions circumscribe discretion, set limits to the disparities that are possible in sentencing similar offenders for similar offences. But some achieve that at too high a price in rigidity.

The mandatory sentence, as the essence of the first revolutionary code in France, was a formula designed to protect against capricious, excessive and arbitrary punishments. It owes its modern reincarnations not to philosophers but to the alarm of public and legislators about the growth of certain forms of crime and the fear that the courts might be going soft on the criminals. The demand for a mandatory sentence of death or of life imprisonment goes up when a policeman is murdered in the course of his duty, when there is an outbreak of brutal robberies, when criminals are believed to be resorting to firearms, when the smuggling and sale of drugs is seen as big business in corruption.

And now, because of the growth of violent crime amongst the young, there is a demand from several quarters in the United States for a substantial mandatory sentence of imprisonment for youths convicted of violent offences for the fourth time. So far as I can judge, Dean Norval Morris, Professor James Wilson and Professor Ernst van den Haag are all prepared to give this proposal serious consideration. The findings of Professor Wolfgang that, in Philadelphia, a small nucleus of recidivists was responsible for two-thirds of the violent offences committed by the young, are being used as an argument in its support. It is contended that a sentence of three or even five years would be no more than such youths deserved. At the same time, by containing them during the most dangerous years of their lives, it would dramatically reduce violence on the streets.

Minimum sentences are a variation on similar themes and spring from the same motives: fear, anger, a sense of defeat, a conviction that an inescapable sentence must surely deter, or at least contain, when all else has failed. Is it not better that offenders should know in advance the rigour of the sentence they will receive? At a lower level, what of the deterrent effect on reckless motorists of knowing that the court has no option but to disqualify them if they are convicted?

What is there to say, then, against mandatory or minimum sentences for selected crimes? Essentially the case is similar to the case against rigid criminal codes: even where they come under the same legal rubric, crimes and criminals are extremely variable. There will be too many hard cases, too many injustices to be swallowed, and therefore too much evasion and twisting of the law by all concerned. When mandatory life imprisonment for selling dangerous drugs was being proposed by Governor Rockefeller in New York, one judge remarked, quite openly, that it was completely impracticable: "District Attorneys, police and judges, would have to figure out some way to get around it." I could back his prediction by many examples from history. It may be necessary to retain a mandatory sentence for murder, because of the very deep feeling that life should pay for life, if not on the scaffold then at least in prison. But even there I remember vividly the difference between the widespread approval of Parliament's decision to retain the death penalty for murder by shooting and the equally widespread horror when it was realised, for example, that this meant hanging a woman like Ruth Ellis, who had shot her faithless lover. The sternly mandatory sentence sounds fine as a general deterrent, quite different as applied to some weak and wretched individual. Yet when modified, as are most sentences of life imprisonment, by the possibility or probability of eventual conditional release, the element of inexorable certainty is weakened and the very justification for a mandatory sentence reduced. Even much lower down the penal scale, the mandatory disqualification of certain driving offenders produces its own anomalies, injustices, difficulties of enforcement, which have to be set in the balance against whatever deterrent potential it may have.

Maximum penalties, upper limits to the punishment a judge may impose for various kinds of crime, are essential to any system which upholds the rule of law. Objections arise only when these penalties are illogical, inconsistent, at odds with people's sense of justice. That can happen when, as in England, the maxima now in force have been laid down at different periods, sometimes widely different in their evaluation of the comparative gravity of various offences and their ideas about severity of punishment. It can happen also under federal constitutions, like that of the United States, where variations between semi-autonomous states add another element of discrepancy. It can happen anywhere when new offences are created and severe new penalties imposed, in response to some particular, possibly transient, panic. It can also happen when an offender is sentenced to serve consecutive terms of imprisonment for multiple crimes. Particularly in some parts of the United States, no maxima are prescribed for these aggregate terms. They have added to the toll of unjustifiable disparities and produced the paradox of sentences extending over several lifetimes.

Thus the problem with maximum penalties is not whether they should be laid down but whether they can be made reasonably proportionate to people's assessment of the comparative gravity of crimes, and a consistent guide to sentencers rather than an additional factor in discrepancies. Many efforts have been made, in many countries, to bring them into line. In England, over a century ago, a body of Criminal Law Commissioners was appointed by the government for that purpose, and worked solidly at the task for over a decade. Now it is part of the work of the Criminal Law Revision Committee. In the United States, where anomalies are sharper and more widespread, lawyers have taken the initiative. The American Law Institute has been the first in modern times to give systematic thought to the principles that should underlie sentencing. Now the American Bar Association has brought out a set of carefully thought out Sentencing Standards, concerned with the legal framework as well as the procedures of sentencing. An unprecedented movement of recodification has swept across the United States and wherever criminal codes are revised one of the main targets is to bring order and consistency into sentencing.

The Model Penal Code of the American Law Institute is a good specimen of the way lawyers have tried to devise a just and consistent scheme of maximum penalties to replace a system, or lack of system, in which maximum sentences for large numbers of offences, differing in different states, are so high as to leave the courts with almost unbridled discretion. It classes as crimes of the first degree only the outrages that appal us—murder, kidnapping, the worst cases of rape and robbery—and allows for these a maximum of life imprisonment. The middle range of offences against the persons of others—less aggravated acts of kidnapping, robbery or rape, together with manslaughter and the graver assaults, would be on a footing with the most serious forms of violence against property, like burglary or arson. But as crimes of the second degree they would carry a much lower maximum penalty than the major outrages: ten years in prison. Other felonies would incur a maximum of five years. And the Code would impose restrictions upon the use and length of consecutive terms.

These maxima may still be considered too high and I would not necessarily endorse them. But, judiciously enforced, they offer a much better solution to the problems of justice and protection of the public than rigid mandatory sentences. They would conform to most people's idea of justice and proportion. Some might think the penal steps between the three classes too steep, or perhaps question whether certain major property offences, like the swindle that ruins thousands, should not incur more than five years in prison. It is one of the major advantages of this kind of approach that it makes such issues clear. It is easy to grasp, easy to challenge, easy to remember, which is more than can be said for the tangle of penalties and gradations that has developed in many modern legal systems. Of course the maxima are by no means

intended to become the norm. Nor need judges resort to imprisonment: indeed in most cases they do not. However, in my opinion there is no doubt at all that the general level of sentencing in
the United States needs to be radically reduced. That is true of
many other parts of the world. Even in England, where the practice is so different, a move in that direction could be made.

On the whole summary courts deal with the less serious
offences, can impose comparatively short terms of imprisonment,
comparatively low levels of fine. In England attempts at further
statutory regulation over the last few decades have been designed
to circumscribe their use of imprisonment still further. Yet they
have not been as effective as was hoped and some, like compulsory suspension of a first prison sentence, did as much harm as
good since magistrates took to using a suspended sentence in cases
where they would previously have tried fine or probation. Now
the compulsion to suspend has had to be withdrawn. There are
limits to the restrictions that can be placed upon discretion without undermining the responsibility, even the dignity, of those
who exercise it. And no measure of restraint can work successfully unless it is reasonably acceptable and in line with the kind
of situation that the courts must face in sentencing. A more
recent proposal in England to take away all power to imprison
from the summary courts and entrust it to the Crown Courts is
unreasonable. Amongst other objections it would increase the
separation between the hearing of the case and the sentence.

Close restrictions upon the use of probation, discharge, suspension, might well eliminate a whole sphere of disparities, real or
apparent. But they do so by cutting down drastically the possibilities of avoiding imprisonment, offering opportunities of rehabilitation where these seem most likely to be taken, tempering
justice with mercy where the circumstances demand it. One of
the greatest penal achievements in England has been the broad
freedom of courts to use these devices, and the extent to which
they have successfully used them. The need is not to restrict this
freedom, but to encourage its fuller use by courts still too much
in the rut of imprisonment. The American Model Penal Code
lays it down that all courts should start with a presumption in
favour of liberty: instead of seeking special reasons that might
justify probation, discharge or some form of suspended sentence,
they should be required to satisfy themselves whether there were

special reasons to justify a sentence of immediate imprisonment. Such reasons would be an undue risk of a further crime during probation or suspension, an offender's own need for treatment that could be given most effectively in a penal institution; or the fact that anything less than imprisonment would depreciate the seriousness of the crime.

Some Americans have proposed that control of the reasoning by which judges arrive at sentence should go much further than that. Judge Marvin Frankel, in a widely read book called 'Criminal Sentences: Law Without Order', demands legislative prescription of the objectives to be sought in sentencing and would stipulate that the judge should be required to state, in each case, which of them—retribution, general deterrence, reformation of the offender, or whatever else—he is trying to achieve by the sentence he has imposed. In addition there would be a list of the factors to be taken into account: it should be possible for the judge to calculate numerically the gravity of the offence before him, he should have clear legal guidance as to whether, or when, he should take account of the prisoner's past record, his personality, background, potential. In other words, the elements and assumptions that lie behind sentencing, and are considered, combined, ignored and weighed, in different ways by different judges in different cases, should be transformed into a fixed legal schedule. Even more simplistic is a proposal advanced by a working-party of American Quakers that the existing tensions and apparent contradictions in sentencing should be resolved by making a clear cut division between attempts to keep offenders out of further trouble by helping them to solve their personal or social problems—the whole spectrum of 'reformation' or 'rehabilitation'—and punishing them by way of retribution and deterrence. Their first few convictions (nature and number to be defined in advance) would bring merely advice on where to find appropriate social guidance. If they failed to accept or benefit from this, and were convicted again, they should thereafter simply be punished on the basis of their offences.

Sentencing is not like baking a cake. You cannot lay down in advance, for every case, the exact ingredients and proportions, the exact temperature required. The further you go in attempting it the greater the risk of wasteful punishment or of outright evasion. There are costs of restrictions as well as costs of discretion. In sentencing courts are not dealing with standard ingredients but with an infinite variety of offences, offenders and situations. Regulation of their discretion has to take account of that. For clarification of principles, a better understanding and balance of objectives, we must look at least as much to those who do the sentencing as to those who set the legal framework.

"Dangerous Offenders"

The crimes that most powerfully arouse a sense of danger are murder, violent assault, robbery, rape, child molestation and kidnapping. An extra dimension is added by the ruthless dedication behind politically motivated violence. And in a different way the violent outbursts by the mentally abnormal stir up a special kind of fear, out of proportion to their frequency, since they tend to be regarded as unprovoked, unpredictable and uncontrollable. It can, of course, justifiably be argued that far greater public danger and personal injury or death is to be anticipated from careless or reckless motorists, or neglectful employers, than from all these categories put together. But the fact remains that these newer transgressions are widely regarded as venial, to be accepted as daily risks inseparable from our way of life. The calculated violence of the professional criminal or political terrorist, the impulsive violence of the mentally abnormal, is not so accepted.

Is there a place for a special kind of sentence, designed to protect society, for a long period of time, from criminals classed as particularly persistent or dangerous?

The year 1908 marked a stage in the development of penal legislation in the English speaking world: the introduction in England of a sentence called preventive detention under which persistent offenders, after serving the period of penal servitude appropriate to their crimes, could be detained for several years more in order to protect the public against their continued depredations. Lord Gladstone, the Home Secretary at the time, and later Winston Churchill when he took over that office, were both at pains to reiterate that this sentence was intended for the dangerous, hardened offender, sophisticated and violent, not for the petty recidivists who were a nuisance rather than a threat. But it did not work out like that: the courts made little use of their powers of preventive detention, and the few offenders caught in the net were just the pathetic, people offenders for whom it had not been designed. So evident were the shortcomings of the system that a committee to examine it was set up in 1932 and it was completely recast after the war as a single rather than a dual sentence, and with stricter criteria of selection. That, too, failed to work and in 1963 a further committee, of which I was a member, recommended yet another version, resulting in the present "extended sentence" for certain persistent offenders. It has been evident for some time that this also is unsatisfactory and is being little used: undoubtedly any review of its application would condemn it in much the same way as its predecessors were condemned. In Australia and Canada, where similar attempts to attack the problem have been made in the past, similar failures and disappointments have been coming to light.

Experience on the Continent of Europe, where the dual system remained the usual model, has been no happier. In theory the second part of the sentence, since it was regarded as purely preventive rather than punitive, was to be served under much less onerous conditions. But a simple old Italian jail-bird, asked what he thought of his move from prison to a detention institution, replied, "I see no difference: the minestrone is just the same." And a solemn Austrian professor characterised the device as an "Etikettenschwindel"—a fraudulent trade description. There is now a move away from this whole conception.

In the United States the main device for extended restraint upon offenders classed as dangerous has, in the past, been the indeterminate sentence. There have been the notorious "sex psychopath" laws and a whole labyrinth of statutes dealing with "habitual offenders." Prolonged and indefinite detention has been justified not only in the name of prevention but of "cure": the offender has been regarded, in one sense, as a patient, to be discharged only when he has responded to treatment and can be regarded as "safe." Nowadays scepticism about the possibility of accurately diagnosing the "dangerous," still more about the possibility of "curing" them, has produced a strong revulsion against such sentences with all their possibilities of injustice, oppression and anxiety.

With all these failures behind us, why is it that, on both sides of the Atlantic, we are once more becoming so concerned with the concept of dangerousness, once more setting out on the search for reliable means of identifying and restraining the dangerous? The Canadian Committee on Corrections which reported in 1969, devoted much of its labours to the dangerous offender. So did the National Commission on Criminal Justice Goals in the United States, which reported in 1973. A more modest, but extremely lucid, report on sentencing and corrections published in South Australia in the same year included a careful analysis of the pros and cons of "habitual offender" legislation. In England, following the horrifying case of multiple poisonings by Graham Young, the Butler Committee reviewed the whole range of provisions for the sentencing, detention and release of mentally abnormal offenders, and now yet another committee has been launched on the specific theme of the dangerous offender. In France, under the stimulus of the Society for Social Defence and of Professor Georges Levasseur of the University of Paris, a fresh exploration has been carried out, conferences have been held and a report written, all on the significance of mental abnormality in criminal conduct and the appropriate response to it.

This renewed concern stems from a variety of sources. There is the element of sheer fear and revulsion provoked by the rise in violent crime. There is a widespread sense of impotence, a pressure that something more should be done to keep out of circulation at least those violent criminals who are caught. Paradoxically, there is also the urgent need to reduce the numbers sent to prison. Many of the American lawyers who advocate much shorter maximum terms for the great bulk of crime and criminals, see the continuance of some kind of extended sentence as a necessary corollary. If the general level is to be far lower, there must be some way of dealing with the exceptional case, of protecting the public for a long time, perhaps permanently, from those whose crimes have shown them to be an enduring menace to the safety of others. The price of using milder measures for the bulk of non-dangerous offenders is seen as devising better means of identifying the genuinely dangerous.

That, of course, brings us back to the crucial question of the criteria of dangerousness, or even of persistence. In England we have felt obliged to make repeated changes during the present

century. In the States the Model Penal Code, the Model Sentencing Act and the National Advisory Commission on Criminal Justice have each put forward different sets of definitions. The
tendency has always been for the criteria to be too wide, too
vague. That makes defence against them difficult. And it fosters
discrepancies in their application by different judges and courts.

We know a great deal more than did the early positivists about
what goes to make a dangerous criminal or a recidivist. We have
studied this in relation to sentencing, parole and after-care. We
can put offenders into "risk categories." But what we are still
unable to do is to distinguish certainly in individual cases, between the person who will commit further crime or violence and
the person who will not. Whether we are trying to forecast
future delinquency in children, the likelihood of further offences
when imposing sentence or granting parole, the tendency is to
overestimate the risks. In the attempt to ensure that special measures of restraint are applied to offenders who pose a serious and
continuing threat, we inevitably bring in others who do not.
That has been the lesson of penal history. And an assessment of
the most up-to-date methods of selection leads to the same
conclusion.

Most of the committees set up to tackle this obdurate problem
have clearly recognised these difficulties. Yet, paradoxically, they
have felt impelled to suggest new criteria, new regimes, new
safeguards, whilst clinging to the idea that a special sentence is
both needed and workable.

All this is not to suggest that there is no such person as a
genuinely and persistently dangerous offender. But the likelihood
is remote that we shall ever be able to pick him out with certainty, whether on the basis of his offence and previous record, of
clinical reports, or of risk categories. Even were such accurate
prognosis possible, it is questionable whether it would be justifiable, in moral or political terms, to make it the basis for prolonged preventive confinement.

The most recent trend in legislation has been to narrow the
scope of such sentences, to ensure that they can be imposed only
where the man's record, his current offence, or both, are really
serious. But where that is the case available maxima are already
high enough in all jurisdictions. I would therefore go further.
Extended sentences should be dispensed with altogether. What is

needed is wiser and more equitable use of their discretion on the
part of sentencers. A man should be punished for the crime he
has committed, not for the crimes he might commit in the future.
If danger is anticipated from his mental abnormality, he should
be dealt with like anyone else whose illness makes him a menace.

The Sentencers

In these irreverent days even judges are judged. Not just individually and sensationally (like Judge Jeffreys with his Bloody
Assizes, 1685) but as an institution and in the course of their
ordinary duties. During the past twenty years, in the United
States, in Canada, in England, in France, the attitudes and behaviour of judges and magistrates have begun to be subjected to
the systematic probings of social scientists. When you consider
how much can depend, for individuals and for society, upon the
way they use their power, it may seem strange that they should,
until so recently, have eluded the microscope. Social standing,
assurance, an air of ultimate moral authority have shielded them
from the impertinent questioner more effectively than prison
walls or official secrets acts have impeded investigation of penal
systems or the police. Yet to understand the response to crime is
as important as to search for its roots, and for that you need
knowledge of both sides of the equation, the judge as well as the
man in the dock. Judges and magistrates are not captive subjects
for research. When, in the eighteen-nineties, there was concern
about the considerable discrepancies in sentences being given by
English judges, Sir James Fitzjames Stephen opposed any investigation, on the ground that it would be an unpardonable intrusion
into decisions which were the personal responsibility of the
judge, a matter for his conscience alone. Today this wholly
uncompromising attitude would not be acceptable. Judges in
many countries are becoming more open. Even so there remains
everywhere considerable distrust of the methods and value of
such investigation, along with the mixed feelings of any profession about having its mysteries probed except by initiates.

What do we expect of those who pronounce sentence? Certainly fairness, freedom from prejudice or bias, from the kinds of attitude that lead, consciously or unconsciously, to improper discrimination. But at the same time we expect them to be aware of a wide range of issues; to be sensitive to the kind of person to be sentenced, what has influenced him in the past and is likely to influence him in the future, and to be aware of the various ways of dealing with him offered by the penal system, and of what each really implies.

Doubts about the ability of sentencers to carry out this double responsibility have clustered round a few main issues. Are they so remote from the majority of those they deal with, in age, in class, in associations, that they can neither judge fairly in terms of culpability nor deal with them effectively in terms of treatment, deterrence or protection? Does political bias, like the racial bias already demonstrated in certain contexts, seriously distort their decisions? Does a legal training help them in their role as sentencers, or does it make them more rigid and legalistic? Do they, at any point in their careers, learn how to sentence at all, let alone learn how to sentence as thoroughly as they learn how to weigh up the evidence of guilt or argue a case for prosecution or defence?

Predictably, the bits and pieces of findings so far available confirm that judicial benches are heavily weighted on the side of the middle and upper-middle class, the middle-aged to elderly, in many places the conservative side in politics, and, with the notable exception of the English lay magistracy, of the legally trained.

Less clear and less uniform is the evidence on how, and how strongly, these features affect their sentencing.

The idea of a judge or magistrate continuing to try and to sentence well into his dotage is horrifying. Short of that the dangers of rigidity, of getting out of touch and of failing to hold public confidence have to be weighed against any benefits of accumulated wisdom. The present retiring age for magistrates in England is seventy, for judges seventy-two or seventy-five. It seems to me that seventy-five is too high. There is still a great deal of room for the appointment of more judges and magistrates from amongst younger people. Yet maturity and experience are of great value on the bench. A study by Professor Hogarth in Ontario showed that it was the bright young magistrates who were more likely to prove extreme in sentencing, whether in the

direction of 'treatment' or of retribution, than their elders who
were more flexible and moderate. French career judges, appointed relatively early, have sometimes been handicapped by
their youth and inexperience. Nor can we underrate the major
service to justice and liberty done by the old lions of the Supreme Court in the United States.

Class influences, likewise, may not affect sentencing to the extent, or in the ways, that might at first sight be expected. That
they exist is undeniable, if for no other reason than that judges
are selected from amongst lawyers. Legal training and the legal
profession have been, at least until very recently, largely the
preserve of the more prosperous classes. England, with three
quarters of its judges, as distinct from its lay magistrates, educated at public schools followed by Oxford or Cambridge, may
represent the extreme. American judges are drawn from more
varied backgrounds, but still come largely from the middle, if not
upper-middle, classes. The bulk of Continental judges likewise
come from good bourgeois homes. Even in the developing countries it is the wealthier families which can afford to send their
bright boys to learn to be lawyers in Europe or America.
Amongst English lay magistrates the bias may not be so extreme,
but it still exists in spite of considerable efforts to find more
working-class justices; even with recent provisions to pay them
for loss of earnings as well as expenses, it is still difficult to find
many willing to undertake the task.

Moreover, the assumption that a judge or magistrate from the
less prosperous classes will have more understanding of offenders
whose social origins are like his own, is not necessarily justified.
Such evidence as there is suggests that working-class magistrates,
like the respectable working class generally, are more punitive
towards people who break the law than are those from easier
circumstances. If you understand more you do not necessarily
forgive more: you may simply feel that, since you have made
good in face of similar obstacles, there is no excuse for those who
have gone wrong. It is possible to suggest that more working-
class judges and magistrates would reduce sentimentality in the
sentencing of offenders. But they are likely to prove a disappoint-
ment to idealists who advocate them in the simple faith that they
will be less likely than others to deal harshly with the lower class
lawbreaker.

If judges are socially remote, it is not only a function of their upbringing. The conventions under which they live and work, especially in England, also set them apart. It is more than a matter of court ceremonial and legal robes. Some of this could well be dispensed with. Though, incidentally, it is wise not to underrate the impact of a 'uniform'. If the young attach so much importance to the length of their hair, their beards, their jeans, their beads, or whatever may be the current fashion, if people turn out in hordes to gaze at the trappings of royalty, why should we assume that it is only the rational that carries weight, that symbols have suddenly lost their potency? But apart from all that, the concerns of judges are not the same as those of other people, even of barristers. There is a sense in which they can talk freely only within their own ranks. The ceremonial and conventions that hedge them around in England contribute to all this, but they do not cause it. I know a judge in New York who had been an enthusiastic poker player, though only in the homes of his friends. He felt it his duty to give up that indulgence when he became a judge, though he admits he misses it. Is that a symptom of social isolation or of a high sense of duty? An American professor said that the very fact that judges there represent a much broader spectrum of background and point of view than in England makes it harder to achieve sentencing based on consensus rather than on personal predilections. The Chairman of the English Bar Council claimed that political health and scrupulous fairness in court could be preserved only if "those who practise advocacy and sit in judgment live and work within the cadre of certain professional disciplines." Can we have it both ways?

At the same time, there is the danger that this apartness from ordinary humanity, the robes, the ceremonial, may minister to a sense of omnipotence and infallibility. Judges occasionally assume the right to harangue as well as to sentence. In most countries there is appeal against conviction if the judge has been unfair in his conduct of the case or his summing up. In some there is appeal against his sentence. Misconduct in other respects, such as rudeness to witnesses, counsel or accused, breaches of courtesy as distinct from breaches of law, may incur criticism in press or parliament, as well as censure within the profession. But some English and American lawyers have argued for more formal complaint procedures.

Then there is the question of politics. The presumption that judges, as distinct from lawyers generally, stand aloof from the mêlée of politics makes it particularly important to keep them as remote as possible from political turmoil. However good the intention, it is a mistake to invoke their prestige in an attempt to quell public rumours, doubts or fears. It may be questioned whether it was wise or useful to invite the Lord Chief Justice to carry out the enquiry into what has come to be called 'Bloody Sunday' in Ireland. The assassination of President Kennedy was an event of a very different kind, but some uneasiness was expressed about the wisdom of asking the Chief Justice of the United States to preside over the commission of enquiry which, for all its formality and thoroughness, was not a trial. Obviously there are the extreme cases, where service to political masters or political objectives are allowed to take precedence over justice itself. Short of that, research on the whole tends to confirm the expectation that the judge who has liberal political convictions, or is a member of a racial or religious minority, will be more likely to favour the defence than his more conservative or conventional colleague. But it is no more than a tendency: many individuals go against the trend.

There are innumerable ways of appointing judges. That local judges should be elected by popular vote was part of the democratic creed of the United States, though it does not affect the Federal system or the Supreme Court. The idea was also adopted for a time in some of the socialist countries.

In the early days in the United States the case for the election of judges was seen in much the same light as the case for the election of legislators: *vox populi, vox dei*. The judges must not be remote, but must represent the values of their local community and be answerable to it. Now that there is a growing uneasiness about the quality of judges, those who still support a system of periodic elections argue that it ensures that judges who turn out badly can be got rid of, and that it avoids the self-perpetuation of a narrow professional establishment. Each of these propositions has been answered in kind. Judges who share the values of their communities can be secured by local appointment without resort to election. To make them too directly answerable to the electors is to risk that they will pander to unjust local prejudices and pressures. They need to be seen as above the turmoil of politics, especially where, as in the United States, municipal struggles are often so tense, violent and acrimonious. If they have to secure election, their chances are likely to depend upon political skill and affiliations rather than suitability for their office. Moreover, in the process of election they may incur embarrassing obligations to their supporters. All these things reduce the dignity of their office, the confidence in their ability to administer disinterested justice. The drawbacks of elected judges have led to various attempts at compromise. In some places, for example, the governor makes appointments from a short list drawn up by the local bar association whilst the electorate retains the right to reject or confirm at the next election.

The English system, whereby judges are appointed on the advice of the Lord Chancellor, himself a judge but also a politician, has worked remarkably well on the whole, largely because of the high standards maintained by the bar. But it has been too much a closed shop. If the legal profession has to be divided into barristers and solicitors, in itself very debatable, they should at least be brought closer together and share the right to be selected for the Bench.

Under the French system, which has been widely copied in many parts of the world, judges start with a professional training, are appointed to the lower courts and work their way up. They hold their appointments under the Minister of Justice and depend upon him for promotion. This also has worked, producing a strong judiciary, though without as much social prestige, perhaps as much sturdy independence, as can be claimed by high court judges in the English-speaking countries.

Inevitably, under all these systems, a political element enters into the appointments. In England it is the Lord Chancellor, a cabinet minister, who has the decisive voice. In the United States it is the President who appoints the federal judges, albeit subject to the approval of the equally political Senate. In France it is the Minister of Justice.

Perhaps the thorniest questions relate to complaints of judicial misbehaviour, to security of tenure and to the possibility of dismissal. We want to have our cake and eat it. We want judges who are too secure to be influenced by fear or favour, whether of politicians or people. We also want some way of disciplining, even removing, judges who abuse their position or who, through infirmity or perversity, can no longer be trusted to administer justice.

Some radical suggestions about this were recently made by a sub-committee of *Justice*, a highly respected body of British lawyers concerned with reforms in the administration of criminal law. Though the governing body rejected their recommendations, they had many wise things to say. They proposed that, in England, there should be a Commission of judges and others, but no politicians, which could advise on judicial appointments, consider well-founded complaints of judicial behaviour, and make recommendations to the Judicial Committee of the Privy Council on whether a judge should be dismissed. Obviously formal procedures of such a kind would raise considerable difficulties. But the idea should not be swept under the carpet and forgotten. At a time when so many old-established institutions are finding that informal networks of control are no longer enough, any reasoned suggestion deserves consideration.

In many parts of the world it is taken for granted that those responsible for sentencing will always be lawyers. That is not the case in England where at least nine sentences out of ten are imposed by lay magistrates. Legal training should certainly help the sentencer in some respects: he should have learned to look at a case from different sides, to select the genuinely relevant issues, to weigh arguments, to reach decisions. On the other hand it is sometimes suggested that it makes him pedantic, detached from real life, more concerned with legal formulae than with genuine human problems. The lay magistrate, by contrast, is pictured as full of robust commonsense, more understanding both of offenders and of the everyday world. To compare them in terms of research is not easy. In England, for example, the lay magistrate is part-time, the stipendiary normally full-time, the laymen sit in groups, the stipendiary sits alone. If there are differences in their attitudes and sentencing, they may stem from such circumstances as these as much as from the fact that the general run of justices of the peace are not lawyers whereas the stipendiary magistrates are. In Canada, however, where the summary courts have all had full-time justices, mostly lawyers but some laymen, a small comparison has been possible. In that situation, at least, it is not the laymen but the lawyers who come out best, as more confident, more likely to take account of a range of factors in the offence, less likely to be rigid in their sentencing, less vulnerable to pressures from those around them.

My own view is that sentencing is not a job for laymen. Such devices as 'people's courts' appear to me highly dangerous. At least in the Western world the moves are toward unified court systems, staffed from top to bottom by qualified lawyers. That has long been the position on the Continent of Europe. It has been recommended, and is in process of adoption, in Canada. It has been introduced in some parts of the United States and there is strong pressure to bring it in elsewhere, especially in cities. There the neglect, squalor and corruption of summary courts, struggling against overwhelming odds, apart from the mainstream of legal protection, legal reform and penal provision in the superior courts, has driven reformers to the conclusion that only a unified system can offer the hope of equal justice and adequate resources at the bottom of the pile.

The English summary courts and the English lay justices seem to me the exception that proves the rule. They have evolved over centuries in the soil and climate of a highly individual nation, part of its strong traditions of local and voluntary responsibilities. Their jurisdiction has been extending rather than contracting even in the most recent decades. They deal with ninety-eight per cent of criminal cases. In terms of expense the saving is immense. In terms of justice and of behaviour there are surprisingly few scandals. In terms of status and respect they still stand high: an invitation to join their ranks is still esteemed an honour. It is a system that does not easily transplant to other countries: it tends either to wither or become corrupt, as has happened in the States. Yet to uproot it whilst it still flourishes in its native soil would be wanton and wasteful. It may be that even there it will in time be starved, eroded by social change. If so it will become necessary to think again.

Almost as vital as the general question of the personal backgrounds and attributes of those given the responsibility for sentencing is the question of their specific training for the task, their opportunities for thinking about it jointly, of pooling experiences and working out guiding principles. It is often pointed out that nothing like as much attention is given to the problem of sentencing as to the problems of ascertaining guilt or innocence. Yet in these days of wide judicial discretion the way the sentence is decided can be as important to the offender and the rest of us, as his conviction or acquittal.

The problem starts right back in the education of lawyers. It is only comparatively recently that criminal law has ceased to be a Cinderella of legal education. Criminology, though fast gaining recognition, is still an optional extra even for those planning to work in the criminal courts. It is not realistic to expect, as some reformers suggest, that judges should be experts in sociology or psychology. Yet some understanding of the factors affecting crime and the penal system is surely essential. So is some direct knowledge of penal institutions.

The organisation of lawyers known as Justice has recommended a Judicial Staff College, which all newly appointed judges would attend for three to six months before embarking on their duties. It is disconcerting that so sensible a proposal was promptly rejected as unnecessary by the Attorney-General. Certainly there have been moves in other directions to give judges more opportunities of thinking about sentencing. The question is whether they are anything like extensive enough.

In the States concern about disparities in sentencing led, as early as 1959, to the launching of Sentencing Institutes by the federal authorities. Within eight years a President's Commission was advocating their adoption by every State. Since 1970 the Lord Chief Justice has been holding similar meetings for English judges. The idea is to bring judges together to discuss the problems and criteria of sentencing. In particular they are given cases raising various issues and each asked to say what sentences he would impose, and why. Comparison of results is illuminating. Apart from the influence of judges upon each other, there is the chance of frank discussion of sentences with experts from outside the judiciary: academics, psychiatrists, probation and prison officers. Some of the most significant shifts in ideas about sentencing amongst judges in the States resulted from information from such people about the legal or practical effects of a particular kind of sentence which a judge may have imposed without appreciating its implications. On the other hand the Conferences also brought out the deep and apparently irreconcilable differences between judges on certain basic issues, especially the deterrent element in sentencing: those most committed to the belief that punishment was a genuine deterrent were likeliest to penalise the defendant convicted after refusing to admit his guilt, to choose prison rather than probation, and, in sentences where parole was admissible, to impose the maximum sentence allowable and reinforce it by a minimum period to be served before release could be considered.

Another American device for persuading judges to make explicit, compare and modify their ideas about sentencing is the Sentencing Council. Whereas the Sentencing Institute brings together judges from many parts of a country or State in a specially convened conference, the Council is simply a meeting of the judges at a single court. The sentences they work out are those they are currently deciding in real life. The ultimate decision rests with the sentencing judge, but there is evidence that, over the course of time, the members of panels tend to approach a common ground, and that in individual cases the sentencing judge may modify the weight he had attached to particular factors in the light of what the others have to say. In fact the whole process is reminiscent of what can happen on a bench of magistrates in England, where justices influence each other both in the short and the long term.

The Sentencing Institute, the Sentencing Council, the much less formal interaction of an English bench of magistrates and the pronouncements on sentencing that flow from Lord Chancellor or Lord Chief Justice, all exemplify the belief that the best form of education in sentencing is mutual education. We must, after all, still confess to great ignorance about what kind of sentences are best for the individual or for society. Differences in outlook and practice have their uses as well as their abuses in this field as much as in, say, medicine or education. We are still frankly feeling our way. Meanwhile, the shock of discovering that a colleague puts a very different interpretation on the gravity of a crime, gives a very different weight to some aspect of sentencing, does something at least to shake ways of thinking or acting that have been accepted with too little scrutiny, followed with too little discrimination.

Supporting Cast

Judges and magistrates do not sentence in a vacuum. Apart from the broader pressures upon them of public opinion, of the trends in crime and penal policy, there are other voices in court which may have a powerful, sometimes a decisive, influence upon the sentence.

The police often complain about leniency on the part of the courts: "We catch them, you let them go." They argue that they know a lot more about what criminals are really like, and how they respond to sentences, than the judge or magistrate, who sees them only in the dock. Should we give them more say? Knowledge of the ways of police states, in which police may be detective, prosecutor, jury, judge and executioner, has led us in England to shrink from any participation of the police in deciding punishment. It is considered that their duty stops at bringing the offender to justice and setting out the evidence against him. In many summary courts in England, however, they also conduct the prosecution. There is some pressure for this 'police advocacy' to be replaced by independent public prosecutors, on the grounds that it is incompatible with their primary duties of vigorous investigation of offences and bringing to justice those they have good reason to suspect.

Quite apart from any role as witnesses or prosecutors, however, the police may affect sentence by what they tell the court of the offender's criminal record, his history, his domestic circumstances and his employment. Some of this information they already have in their records, some of it they may pick up in investigating the crime, some of it is sought expressly as additional information for the court. Whereas most probation officers in England would have scruples, except with the express agreement of the defendant, about approaching an employer before a conviction to check his work record, English police assume the right to do so as a matter of course when a person is committed for trial. Again, English police have in the past used the opportunity of reporting on antecedents to comment rather freely on such matters as the criminal associations of the accused, his bad character or way of life, or the trouble he had given them. This has been firmly cut down by the stipulation that such adverse statements must not be made unless they can be supported by factual evidence. There may be plenty of gossip and complaints about a man, but it is not so easy to get people to come and give evidence about them in court. It still remains open to the police to tell the court that an offender has been co-operative in the course of their enquiries, with the implication that allowance for this should mitigate his sentence. Even that has its dangers: more than one defendant has complained that he admitted to things he had not done because the police said they would 'put in a good word for him at court.'

What of counsel for the prosecution and defence? If the 'adversary process,' whereby each side tries to make as strong a case as possible, is the best way of bringing out the evidence on guilt, may it not be the best way of arriving at a balance about sentence? They have gone much further along these lines in the United States than in England. The English view is that the prosecution, like the jury, should confine itself to the issue of guilt, keep clear of considerations of sentence. At most counsel occasionally comment that the offence is becoming very prevalent, is a particular menace to society. In the United States, as in many countries on the Continent of Europe and of course in all socialist states, there are none of these inhibitions. The prosecutor is expected to demand severity if he thinks the case calls for it and does not hesitate to give his view on what the sentence should be. The circle is completed by defence counsel, who will try to get a sentence that will be in the best interests of their clients, almost always interpreted as the most lenient possible in the circumstances.

The result of allowing the battle between prosecution and defence to continue into sentencing has been to infect the earlier stages of investigation and prosecution. The sentence, in short, can become a major bargaining counter in securing admission of guilt. In many States, though not all, sentence is virtually decided by this means before the case comes to the court. The dropping of other charges against Vice-President Agnew, with a sentence of no more than fine and probation in exchange for his agreement to save time and embarrassment all round by not contesting a charge of tax evasion, was a spectacular example. But it was far from being an exceptional device for dealing with an eminent person. In New York, ninety per cent of all criminal cases are settled in this way. In many other cities too it is the device that keeps criminal justice moving. The interest of the prosecution in securing a plea of guilty, especially where the evidence is thin or cumbersome; the interest of the defence in keeping the charges made and the sentence imposed as low as possible; the interest of the court, especially when crime is high, the pressure almost intolerable and delays accumulating, in the avoidance of lengthy trials by jury wherever possible: all these converge to foster so-called 'plea-bargaining.' Though the proportion of guilty pleas is much the same in countries that do not rely upon plea-bargaining as in those that depend upon it, there are many judges, public prosecutors and other practising lawyers who still believe that it is indispensable, that the whole system of justice would break down without it.

In England the practice of plea-bargaining has, until recently, been far less openly admitted or widely discussed. Bargaining undoubtedly goes on under the counter between police and defence. Though the issue is the gravity and number of charges to be brought, this can, in itself, profoundly affect the sentence. Now it has been conceded that it is also accepted practice for a judge to give a shorter prison sentence (though not to change the nature of the sentence) where there is a plea of guilty. In 1968 the Court of Appeal reduced a sentence by a third because no discount had been allowed for the plea. On the other hand, though the judge may informally indicate to council what reduction he might allow in a particular case, it is considered improper for counsel to pass this on to the defendant, except under the guise of his own professional opinion. That is no doubt because, if conveyed as a message from the judge, it could be interpreted as direct pressure from him. In 1970 a motorist appealed on the ground that he had pleaded guilty only because his counsel told him he would otherwise risk imprisonment. The Lord Chief Justice stated that he was sure the defendant had been mistaken in believing that this warning came from the judge, but he felt it necessary to add that a judge should never indicate what sentence he would impose in the event of a guilty plea lest he gave the defendant the impression that, by pleading not guilty, he would risk something worse. Pressures on the courts and delays in justice are not nearly as serious in England as in the United States, but they certainly exist. Plea-bargaining is incomparably less extensive, its impact on sentences far less decisive, but again it exists.

The inevitable result of allowing plea-bargaining, however carefully wrapped up, is to put heavy pressure on defendants to plead guilty, whether they are so or not. Suspects are told that they should not be influenced by any threat or promise in making admissions. Is it any better if the warning is labelled an inducement rather than a threat and comes, however indirectly, from the judge rather than the police?

Sentences reached by such means depend less upon justice or the needs of the offender or of society than upon the skill and weight of the negotiating counsel. If anyone benefits it is the wealthy offender who can afford first class legal advice from the outset, and the recidivist who knows the ropes: how long to hold out, when, and at what price, to clinch the deal. Inevitably such a system makes a massive contribution to disparities in sentencing, and must destroy any faith a defendant may have in the fairness of the system of criminal justice. Whilst some would-be reformers argue simply that the system should be regulated, brought into the open, others have taken the line that it is fundamentally mischievous and should be outlawed.

Prosecutors must have a measure of discretion, indeed they have it in every country, but it is another matter to allow them to become the arbiters of sentencing. Their bargaining powers were never intended to assume the nature and dimensions they have now attained in the United States. It would be a rewarding study to see how the practice has crept into criminal justice since the beginning of this century, at first gradually, but at an ever accelerating pace, until it has attained the status of a system within the system. The sooner it can be abolished the better, but that is easier said than done. The fact that it exists on so vast a scale reflects the crisis through which the American system of criminal justice is passing. As long as the present pressures continue it will continue too.

There has long been a trend towards assuming that sentencing can be improved by fuller information about the offender. In England we have witnessed more frequent use of reports from probation officers in courts at all levels. In the States such reports are the usual rule in the Federal system and are widely required in certain States; moreover pressure is building for many more to be available in summary courts, where they are at present rare or nonexistent. Even in France, where information about the defendant has hitherto depended largely on the police and on what the magistrate can ascertain direct, the suggestion has begun to be heard that probation officers should be recruited and trained to make enquiries. But to accept that reports are desirable is not to say that they should be immune from criticism or scrutiny. On the contrary, the greater the part they play in influencing sentence, the greater the need to ask when and where they are needed, whether their contents are reliable and relevant, what protection there should be against mistakes, misunderstandings or abuse. Recently there have been those who have questioned the justification and relevance of reports from probation officers and psychiatrists on the personality and background of offenders, arguing that all we have a right to do is to punish him for his crime, all that we have a right to know is the circumstances of that crime.

No one seriously suggests that courts should have full reports on all the streams of minor offenders who come before them, mostly to be dealt with by fine or discharge. Even so, both in England and the States, they have the right to ask for reports upon anyone found guilty: there is, for example, the person whose mini-assault may be the first symptom of far more serious trouble. Beyond that, there have been various attempts to specify circumstances in which courts should always have reports before they sentence: offences serious enough to incur more than a year or two in prison; resort to measures like probation that depend a great deal on the personality, situation and response of the offenders; people who need to be given special consideration, because of their youth, or because of their previous good records, or because they show signs of mental abnormality; people who have never been to prison before but may now have to go there for the first time. In England, so far, courts have been recommended not to proceed without considering reports on certain categories of this kind. In the United States it is left, on the whole, to the courts to decide whether they want information, though in a few States they are obliged to get a report before using probation. The American Bar Association, in a thorough review of sentencing practices, has suggested that the need for a report is strongest where a person may be deprived of his liberty for a year or more, where he is a first offender or where he is under twenty-one, but that the use of probation could be discouraged if reports were made mandatory, especially where courts are hard pressed and have too few officers.

There has not, so far, been much controversy about the right of courts to ask probation officers or psychiatrists to investigate the lives and personalities of the people they have to sentence. There has, however, been dispute as to whether such enquiries are justified in advance of trial and conviction. A full-blown investigation takes time, probably at least two weeks: the defendant has to be interviewed, perhaps repeatedly, often his family, sometimes his employer as well, and there may be the need to check. But an adjournment for reports after conviction is a hardship for the defendant, an extra burden on the court. So in England it has been accepted that, unless the accused objects, the investigation can take place before the case is heard. In practice objections are rare. The American Bar Association would reverse the emphasis: investigation before adjudication of guilt should be allowed only if the accused consents and only with the proviso that the report shall not come to the attention of the prosecution or the court before guilt is decided. That is the position in English magistrates' courts but the judges in the higher courts receive a copy of the report, along with the criminal record, before the case is heard. The argument for this has been that they cannot be influenced in the decision on guilt, since that is a matter for the jury, and that their judicial impartiality is an adequate safeguard against any danger that the report may influence their summing up. There has, nevertheless, recently been mounting opposition to this amongst probation officers. In spite of the inconvenience, even hardships, of deferring sentence to allow judges to consider reports and criminal records, there is a strong case for withholding these until after conviction. It may also happen that, with growing sensitivity about invasion of privacy, the practice of carrying out investigations in advance of trial, whether by police or probation officers, may become less generally accepted.

If the probation officer is licensed to investigate, his investigation should be confined to what is relevant to the sentence. The trouble is that no one is yet sure where the limits lie, and what is relevant in one case may be irrelevant or marginal in another. By training, officers are deeply interested in family history and relationships. By experience they regard the records of crime on one hand and of employment on the other as crucial. They are concerned with trying to assess attitudes, and especially whether a person would respond to probation if that is on the cards. In all this they must inevitably select, weigh up information, on the basis of their training, experience, intuition. There is evidence that their approaches and choices are no more uniform than those of the judges. A few criminologists have suggested that their reports should be informed, supplemented, even replaced, by tables showing the risks of reconviction in relation to a few very simple factors in an offender's history, like his previous criminal record. Such tables are very useful in research. But they have certainly not reached the point where they offer much help in deciding the sentence for an individual offender.

Should a probation officer stop at selecting and evaluating information about the defendant? Or should he go a step further in influencing sentence and give his opinion on whether some particular measure would be more likely than others to keep the offender out of trouble in future? With the encouragement of the courts, he often does so, especially when the crucial decision is whether or not a person will respond to probation. Sometimes he will suggest outright that a fine or a prison sentence would be more appropriate. An investigation of what happened when probation officers in the Federal Courts of the United States made recommendations for or against probation indicated that nine out of ten of those recommended for probation were given it, as against only two out of ten of those considered unsuitable by the officers. The level of agreement was much the same in all the courts, in spite of the fact that some used probation a great deal, some comparatively little. Of course, it might simply be that the probation officers, rather than influencing the sentence, were tailoring their reports to the views of their judges, to what they felt the courts would swallow. But evidence of considerable differences in the recommendations of different officers within the same court goes against that hypothesis.

Particularly where judges or magistrates have a hand in appointing probation officers, they tend in general to respect and follow their advice. Sentencing is a lonely and responsible job. Many magistrates and some judges are glad to have someone to share the onus. On the other hand, they select the information they will take into account, the opinions they will follow. A judge who has determined that, in the cause of retribution and general deterrence, a woman who has persisted in fraud must go to prison, will give short shrift to a probation officer who tactlessly presses upon his attention the fact that she has several small children. On the other hand there are the cases where courts put people on probation against the judgment of the probation officer, and the rate of failure in such cases is not so very much higher than amongst the rest.

Doctors or psychiatrists may similarly be seen partly as experts called in to help the court decide upon sentence, partly as people with whom the sentencer can share a measure of his responsibility. Their opinion may be decisive, especially where there is evidence of specific mental illness related to the crime and calling for hospital treatment. A diagnosis of 'psychopathology' is a much vaguer affair, since the experts themselves disagree both on the meanings and the appropriate treatment. Or a psychiatrist may merely be using his professional authority to repeat and reinforce the commonsense opinion of police or probation officers that a particular offender's problems are predominantly social and that medicine has little or nothing to offer. Judges and magistrates vary in the faith or scepticism with which they regard psychiatric advice. There can be grave dangers in ignoring it. But both Russia and the United States can furnish examples of the dangers of slavishly following it. Psychiatrists cannot X-ray the human disposition in the way other doctors can X-ray the human body: they are not infallible in their diagnosis, prognosis or prescription. Yet what they say touches on the raw nerve of fear of the abnormal and the instinct to take no risks with the 'mad.'

If the reports of probation officers or psychiatrists can so influence the decision about a man's fate, including the crucial issue of his liberty, should they not be as hedged around with safeguards as the evidence that can lead to his conviction or acquittal? Inevitably they are partly based on hearsay, on the opinions of others, which may be biased, on the judgment of the person making the report. Inevitably they are selective: they cannot pretend to give the 'whole truth.' They may reflect personal bias, or misunderstandings, misguiding rather than guiding the sentencer. The right of the defence to see and challenge the report of a probation officer is established in English law, though, for an ill-educated and undefended offender that may be of little help unless the vital points are also explained to him. In the United States the right to see the report has been the subject of much heated debate. In most States it is left to the discretion of the courts and in practice the trend is towards more disclosure, except for confidential matters. In only a few States, however, is disclosure specifically required.

What arguments can there be against so basic a right? Some of those advanced seem to reflect expectation of more violent reactions than have occurred in England. There is the concern lest, without promise of secrecy, sources of information might dry up from fear of court proceedings, local feuds, rows within families or amongst employees. There is the suggestion that there would be endless challenges, delays and demands that witnesses be brought. There is the argument that there is nothing unfair to the defendant in withholding the reports from him since, as it has been put, "after conviction a case ceases to be an action at law and becomes a social problem." Looked at squarely, if reports are collected in such a way, their contents of such a nature, as to provoke this kind of reaction from their subjects, there seems all the more reason for allowing the opportunity to question them. And if sentencing is a social problem it is all the more necessary to let the person most involved know how it appears to those who plan to solve it. Even apart from any risk that errors or injustices have slipped in, account must be taken of the prisoner's sense of fairness. There is always the fear that things said behind your back have been to your detriment.

But what if the contents of the report are indeed to the detriment of the offender, whether in the short or the longer view? A wife or child might genuinely be threatened by violence for speaking frankly. The outlook for the offender himself may be gloomy, perhaps because of progressive mental or physical disease. The probation officer, who may have to supervise him if he is put on probation, may have adverse comments to make on his behaviour and may even forecast that he will fail on probation. Yet it is absurd to say that the whole mass of offenders should be denied the right to see reports lest harm should be done to the handful who might be damaged by them. Surely, in all the exceptional cases imaginable, there should be legal representation at the time of sentence, defending counsel should see the whole report and only that part be withheld from the offender that might do him harm.

A further claim is sometimes made that reports should be fully public, read out in court for all to hear. Reports contribute to sentencing, sentencing is part of justice, runs the argument, and justice should be seen to be done. But that would be to compound the invasion of privacy. Enquiries relevant to sentence may be justified as in the interests of society and the offender. Public exposure of details of his private life, and that of those close to him, goes far beyond that: it can be an additional penalty, for him and for his family, a formidable obstacle to his restoration to normal life. The best remedy for public doubts about the reasons for a sentence is a reference to those reasons by the judge or magistrate who imposes it.

Correcting the Sentence: Appeal

Because judges and their advisers are variable and fallible there will always be some injustices and miscalculations in sentencing.

If you think you have been unfairly sentenced in an English summary court you can appeal to the court above, the Crown Court. Your case will be heard again and your sentence may be lower, the same or more severe. So there is a risk. If you are sentenced in a Crown Court, your appeal is to the Criminal Division of the Court of Appeals. The judges there will consider the principles of your sentence. If they think you should have been sentenced differently they can replace the decision. If they consider your prison sentence is seriously out of proportion they can reduce it. If they think that, though you deserve the sentence of prison, your case warrants a chance on probation, they can make a probation order. If you have been sent to borstal for training, and they think probation equally likely to succeed, they can again order probation instead.

All these constitute reductions in sentence. Under the existing English system the appeal machinery dealing with these more serious Crown Court cases has no power to increase the sentence on an appeal by the defendant. Nor has the prosecution any right of appeal against a sentence in any court which it considers too lenient. This raises the question of whether the system is tilted too far in favour of the criminal. Should he not have to run a risk of a stiffer penalty if he appeals on the ground that his sentence was too harsh? It has been pointed out, with some asperity, that under the existing system the judge or magistrate whose sentences are too mild can pride himself on never having been reversed on appeal. And it has been suggested that the risk of a severer sentence would deter people from wasting the time of the courts by putting forward frivolous appeals.

It is significant that, until 1966, the Court of Appeal had the power to raise the sentence but they did so very seldom. Moreover the danger of frightening off people who have a genuine case for reconsideration has been held to outweigh any advantage. Though the power to increase sentence has been abolished, deterrents against unjustified appeals remain. To start with, leave to appeal has to be obtained from a judge, and may be refused; and it is now the rule that time spent in prison pending an appeal does not count against sentence unless the court orders otherwise. But it has to be admitted that removal of the risk of higher sentences and greater availability of legal aid for appeals has, on balance, considerably increased their frequency. You can see this as an inevitable cost of giving everyone full access to the remedies against injustice, or you can see it as an intolerable addition to the burden of the courts at a time when crime has been getting out of hand.

In the United States this last consideration has been one of the major obstacles to the general establishment of a system of appeals against sentence. Just as there is the feeling that the courts would be overwhelmed by not-guilty pleas if plea-bargaining were outlawed, there is the fear that the judges would be overwhelmed by frivolous appeals if appeals against sentence were widely adopted. Yet in the absence of machinery for such appeals an enormous amount of court time and legal ingenuity is squandered on appeals against conviction. If you cannot get your sentence reduced any other way, the only solution is to find some technical error in the hearing of your case. The situation is further complicated by plea-bargaining: how do you stand as regards appeal if you get the sentence your counsel negotiated for you? A man is entitled to a fair trial and sentence. He should also have a right of appeal against a sentence that may be unduly harsh. It should not be impossible to devise ways of sifting appeals so that only those of substance are passed on for full consideration by tribunals. Arbitrary refusal of the right to appeal, or its limitation to cases involving a substantial term of imprisonment, is a denial of justice.

Closely tied up with the right to appeal against sentence is the question of whether courts should be required to state their reasons for a sentence: for example, if it involves long confinement; is its primary purpose to give the offender what he deserves, to deter him or others, to reform him, or to keep him out of harm's way? If that is stated outright the prisoner knows what he has to appeal against. It is perhaps partly on the principle of least said soonest mended that judges, in general, are strongly opposed to a statutory requirement that they give a formal statement of reasons when sentencing, though they may often do so spontaneously. Sometimes there may be the same kind of fears of causing future harm that lead to claims that probation or psychiatric reports should be secret. Most basic of all, however, can be the reflection that a long sentence may simultaneously serve any or all of the purposes of punishment: why compromise any of them by specifying? A judge who feels like this may take refuge in stock phrases and clarify little.

Appeal courts have the function not only of remedying individual injustices but of laying down principles and precedents which all lower courts are expected to follow. That means that they too need to give reasons for their decisions and that the reasons need to be publicised. In England, it has been claimed, the judges hearing appeals against sentence have in fact built up a reasonably coherent system of principles, though mainly by oral tradition. Even legal journals have, until recently, tended to report only an occasional striking decision. It is still more recently that first attempts have been made to bring them together and analyse their reasoning.

Replacing the Sentencers

Why struggle to repair, renovate, improve the ways courts sentence people? Why not just take the job away from the judges and hand it over to specialists? Institutions, as well as equipment, are surely expendable nowadays. True, sentencing has been the prerogative of the courts since the old, bad days when it was wrested from the tyrants. But is that any reason why the judges should hang on to it now that times have changed and everything is a job for experts?

The idea is not as novel as it may sound. The courts have already parted with a large measure of their control over the punishment and treatment of offenders in several ways. In many countries minor offences, especially traffic offences, are dealt with by fixed fines, imposed at the discretion of the police. Taxation authorities, we recall with a shiver, can exact their own retribution for evasion. Administrative tribunals can impose penalties for certain commercial offences. Professional bodies, and indeed industrial firms, impose their own punishments upon their erring members. "The length of time a child can be kept in an approved school or community home will no longer be governed by his offence or a court order. He will stay there as long as we think necessary for his treatment or education." That was the point of view of an English children's officer, anticipating the act passed in 1969. The system would be completely flexible. The child would be observed, helped, understood. As soon as he was ready he would be allowed to go home. But how long might he be held if he failed to measure up? The treatment of young offenders has served as a spearhead in a more general onslaught on the powers of the courts. In Scandinavia, in parts of the United States, and now in England and Scotland, the theme has been that decisions about the treatment of a child offender should be taken out of the criminal courts, handed over wholly to the discretion of administrative bodies concerned with his education and care.

Even with adults the power of the court to control the length, nature and security of confinement has been greatly diminished, over the present century, in some cases removed. In the old days the court could decide between 'hard labour' and the less rigorous regime of ordinary imprisonment. Now it has no voice in the choice between an open prison or a closed one, trade training or psychiatric treatment, release direct or via a prison hostel, under supervision or without it. If an English court sentences a youth to borstal it has no influence on how long he stays there; the outside limit of three years is set by law, the actual date of release is determined by those in charge of him. If it sentences a man to a substantial term in prison there is no guarantee that he will serve more than a third of it: that is a matter for the parole board to decide.

The idea that criminals, at least the more serious and persistent of them, should be dealt with on the basis of their individual dangerousness to the community and the possibility of reforming them can be traced back at least to the Italian Positivists. If you believed you could infallibly identify the 'born criminal,' who would go on robbing and murdering if left at liberty, it was natural enough to suggest that his detention should depend on his criminal prognosis rather than his immediate offence. Similarly, the offender diagnosed as likely to respond to correction should be held as long as it was necessary to make him safe, whether or not that period corresponded with the gravity of his crime. The theme was taken up and further developed in the nineteen-twenties. A criminologist who was devoting his life to the search for constitutional and social peculiarities that would clearly distinguish between persistent delinquents and the rest of us, claimed that sentencing should be handed over to boards of experts—sociologists, psychologists, psychiatrists, social workers—who would be in a position to do the job scientifically, to relate the sentence to the reformation of the offender and the protection of others rather than to the old-fashioned ideas of retribution cherished by the courts.

A very modified version of this idea would leave the courts to deal with the general run of offenders but pass to a sentencing council the most awkward or doubtful cases. The council would decide what to do with the feckless drifter, who keeps on thieving undeterred by fine, prison or probation, who has far less than the normal ability to cope with modern life and who cannot be expected to respond to psychiatric treatment. It would weigh up the case for extended detention where a criminal might be expected to be too dangerous to release after the normal sentence for his crime. It might also take on the job of deciding whether a particular offender would be suitable for some new and experimental measure.

The case for separate sentencing tribunals rests on questionable assumptions. It is not yet true, if it ever will be, that offenders can be divided into clear-cut categories. Nor is it true that types of treatment appropriate to each category have been distinguished, let alone made available. Nor yet is it true that experts from various disciplines, even experts within a single discipline, will necessarily agree about diagnosis, prognosis, treatment. You have only to look at the disputes raging in all branches of the social sciences to know that. Nor do 'experts' escape the personal variations in attitudes and philosophy blamed for some of the disparities in judicial sentencing.

On the purely practical side the difficulties are formidable. Is a time of rising crime, delays in justice and strain upon the whole system a time to introduce yet another stage, yet another delay, into the criminal process? What of the assumption that a sentencing board will have far more information about the offender than the courts can have, perhaps delay sentence until it has watched his progress in custody, perhaps alter it as it goes along? Courts have already the power to remand for reports from experts of various kinds. To allow more delay than that in telling a man his sentence creates dangers of its own. Then there is the great shortage of experts; what hope is there that enough of them could give sufficient time to his task of sentencing to become more experienced in it than the judge?

One way and another, no country has yet tried sentencing tribunals as alternatives to the courts. But some have gone a good way towards it by the combination of so-called 'indeterminate sentences' with 'time fixing' by administrative boards. California started on this road at the end of the nineteenth century, albeit impelled less by positivist theories than by distaste for the wide disparities in the sentences imposed by her judges. There is still considerable statutory authority for a court to sentence to a jail term of up to a year, but beyond that the position seems to be that little real choice is left to the judge. If he imposes a prison sentence it must be for the maximum term laid down by statute: robbery, for instance, carries a maximum of life, with minimum of five years to be served before parole can be considered. The judge cannot alter this, cannot take account of mitigating circumstances. His sentence is, in this sense, a formality. It falls to the 'Adult Authority' to fix the actual term to be served, the date and duration of parole. George Jackson, the 'Soledad Brother', was sentenced at the age of nineteen to the statutory term of 'five years to life' for armed robbery. He first became eligible to be considered for parole in 1962. By 1970 he had been refused parole ten times, had accumulated forty-seven offences against discipline on his record and was charged with the murder of one of the guards. He had never had a release date to look forward to and was never released alive.

Judicial sentencing has been going on for a long time. There has been ample opportunity to demonstrate and analyse its defects. Sentencing by experts on the criteria of dangerousness or reformation has had a much shorter and less public trial. The Californian experience, which is the longest, is not encouraging. A prisoner may feel aggrieved if he thinks a judge has sentenced him unfairly: the best remedy for that is appeal. But a prisoner under the Californian system, who may have to wait for nearly a year before he knows how long he must serve before being considered for parole, can be in an even worse predicament. He may feel he is being punished for not conforming in prison rather than for his original crime, believe that his release may be delayed by way of example to other prisoners. And there is no appeal except to the authority whose agent has already refused parole. The whole process of determining how long a man shall serve is essentially private. The safeguards of publicity and due process are absent. No reasons are given for refusal. Abuse and the suspicion of abuse is almost inevitable. Yet legal safeguards are rejected on the ground that all this is not 'sentencing' but only 'fixing the term.' The ideal may have been to reduce unfairness, make the most of chances to reform, avoid unnecessarily prolonged detention. But in practice the system has created new injustices of its own, new obstacles to reform, even prolonged detention. Moves to take the whole substance of sentencing out of the courts have been pushed to the point where it is moving not forward but backward, circling back round towards the old horrors of arbitrary confinement.

It may be said that the function of the judge is not to make the law or constitution but to interpret it. Yet the interpretations themselves become part of the pattern of justice. This is especially true of sentencing, where the latitude is widest. But the position of the judge in a settled society, with homogeneous values and little change in the air, is far less demanding than his predicament in times of swift change and of internal tensions. If he works within a balanced democratic system, capable of adjusting to change, he can help to hold the balance true. English judges have been criticised for the severity of their sentences on hooligans making racial attacks in Notting Hill, on students smashing up an hotel as a left-wing protest in Cambridge, on the conspirators of the Angry Brigade. But beyond the criticism was acceptance that they were not politically motivated, they had punished unacceptable violence from whatever class or quarter.

Where tension becomes terrific, passing beyond sporadic outbursts towards a violent upheaval, judges can no longer hold a balance. Do they then concentrate upon shoring up the existing system, or do they take arms themselves on the side of change? The answer is not as simple or single as might be expected. Much depends on judicial tradition, status and independence in the country concerned.

The most obvious reaction is for the judge to support the existing regime. At lowest he may depend on it for his appointment and tenure: it represents his own way of life, even consists of his friends and relations. At best he may be convinced that his whole duty is to enforce the law as he finds it; the law embodies the will of those in power, so he has no choice but to support them. In Spain, in South America, wherever judges are no more than nominees of the regime, they will, when it reaches the crunch, come down heavily against those who threaten it. Severity against political offenders spills over into more severity against ordinary criminals as well. There is little concern to waste on abuses in procedure, even police torture as a means to extort confessions and information. The overriding need to maintain the security of the regime covers a multitude of sins. Yet even when so deeply implicated, judges dislike open political involvement, are relieved if the task of dealing with political offenders is taken out of their hands by military courts, although the shifting of the onus to others may still further reduce the quality of justice, the respect for the courts.

In France and Italy, to some extent even in South Africa, you have, instead of monolithic support for the existing regime, a tug-of-war between judges of different outlook. In France it largely reflects the gap between generations. The younger are for change, the old for the established regime. The younger judges publish their own periodical, voicing outspoken criticisms of their elders. This questioning, even conflict, is healthy up to a point. But if the split continues too long it becomes unhealthy, dangerous. What respect can people have for justice if they hear judges contradicting and accusing each other?

There have been times of upheaval when the judges have neither supported a despot nor split into warring camps but drawn together to defend what they see as justice, even against the regime in power. No doubt they have also been asserting or defending their own independence, but they have done so at the risk of losing office, sometimes even life. Before embarking upon their more monstrous solutions, the Nazis had to undermine or by-pass the judiciary. *Ad hoc* devices ranging from Star Chambers to revolutionary tribunals bear witness to the obduracy of a country's established judiciary where its political leaders go too far in demands on their allegiance. The United States explicitly entrusts to a body of judges, the Supreme Court, the task of interpreting and safeguarding its constitution. It is in extreme situations like these that the judges come nearest to a Platonic ideal of justice, are least bound by national limitations.

When tension has cumulated in violent change and a new regime is firmly in the saddle, the judges of a country face much the same problem as foreign governments. They may disapprove, they may be reluctant to jetison past loyalties, but they are forced to recognise the reality of power if they want to retain any foothold, any chance to moderate action or to protect the rights of those for whom they are responsible.

Judges are expected to be superhuman, sitting like God Almighty above the tumult of the world, perceiving all things, weighing all things, coming in at the end of the criminal drama to put everything to rights, in short doing justice. Witness the way we call upon them, as final arbiters, to give the binding verdict on the public scandal, the trade dispute, the constitutional impasse. Witness *habeas corpus*, not just a legal device to secure the appearance of a prisoner before a judge but a symbol of a free society. We rely upon judges as guardians of the nation's integ-

rity as well as its order. We assume in them a purity of conscience as well as of reason. We expect them to defend us against
oppression even by our own government or its servants. Justice
and judges, unlike law and order and policemen, are required not
merely to be useful and efficient but to uphold some absolute,
impartial, even transcendental ideal. In England their dress and
the ceremonial which surrounds them reinforce their symbolic,
almost sacred, status. In every country they are closely associated
with the national flag, the national ideals.

Yet that in itself is a reminder that judges do not simply embody some Platonic ideal of justice. They are human beings,
breathing the air of a particular country, sharing its distinctive
values. Their assessments of what is just or reasonable, merciful
or possible, are not made in a void. They depend on the outlook,
the social and economic arrangements, the penal or restorative
devices, in the midst of which they live and have been reared. In
certain nations they breathe the air of primitive autocracy: it is
nothing to send the executioner down to the prison to take off a
man's head. In others they breathe the air of totalitarian conviction: the deviant who challenges the faith must be changed,
exiled or eliminated. In others they share the fears of wealthy and
entrenched capitalism: the communist, the revolutionary is to be
suppressed more ruthlessly even than the common criminal. The
fierce democratic insistence on the independence of judges is
evidence of the faith reposed in them. But what they defend is
not absolute righteousness. It is the law and constitution of a
particular state, whether it be England or South Africa, Israel or
Libya, the United States or Brazil. Even at their most magnifi-
cent, when they defy what they see as abuse of power in high
places, they resist it as an infringement of the ideals of justice, the
rights of citizens, enshrined in the traditions of their own nations.

Part V

THE PENAL
PREDICAMENT

CHAPTER
9

Prisons in the Pillory

A Time of Disillusion

Today it is scarcely possible to mention prisons without bringing in the word "crisis." Recidivism, overcrowding, protest, have contributed to an atmosphere of mounting disillusion, even despair, about penal institutions.

This is in sharp contrast with the optimism that prevailed just before and after the second world war. In England during the nineteen-thirties borstal training for the more serious young offenders, with its system of classification, specialised institutions, and after-care, could claim to be succeeding with seven out of ten of its graduates. Now the proportions have been reversed: almost seven out of ten are known to revert to crime within two years after their release. If that has been the fate of the institutions most specifically designed and run with an eye to re-education and rehabilitation, what can be hoped for the rest? Over one half of the less hardened young offenders sent to detention centres are similarly reconvicted. It is a sorry story all round.

In 1936 there were less than eleven thousand people in the prisons of England and Wales; a smaller segment of the local population than in almost any other country in the world. And the number was decreasing. Every man had his own cell. The conditions of ordinary life outside were improving. There seemed no reason why the principles apparently so successful in dealing with the young in borstals should not, in due course, be extended to prisons for adults. After the war, in the Criminal Justice Act of 1948, the projects and promises of the nineteen-thirties were taken up. Administrative changes forged ahead. There were open borstals and open prisons. Prison welfare and after-care were reviewed and re-reviewed, and eventually put into the hands of probation officers, trained as social workers. There were various experiments designed to involve prison officers more fully in rehabilitation as well as custody. Prison hostels, or arrangements for employment outside, were introduced to pave the way for release for long-term prisoners. None of this, however, availed to stem the tide of recidivism.

Now virtually all the features that favoured prison reform and experiment in the thirties have been reversed. Few prisons have any longer the elbow-room for improvement. They are crowded beyond belief, with nearly four times as many to hold as they had then. England can no longer boast of the low proportion of her citizens in confinement: it is now higher than anywhere in Western Europe except Germany and Austria. For every hundred thousand inhabitants the United Kingdom has eighty-two people in prison, whereas Belgium, France and Denmark have brought their quotas down to between fifty and sixty. Well over a quarter of all prisoners—at least twelve thousand—have to share cells with one or two others. And the total prison population is going up, not down—it stands at over forty thousand now and we have been variously warned to expect fifty, sixty, seventy thousand within the next decade. Outside conditions are less promising than they were: with shortage of work and shortage of housing the old jealousies of attempts to improve the lot of prisoners are easily resurrected. And inside, though England has so far escaped the kind of major violence that has occurred elsewhere, there have been new kinds of unrest and demonstrations.

To find comparative statistics on such matters as the numbers of people imprisoned, recidivism, prison disturbances or disciplinary measures, is even more difficult than to find reliable statistics of crime. These are not matters which any nation or administration wishes to advertise. A recent United Nations census showed the United States as having the highest proportion of its citizens in custody: one hundred and eighty-nine out of every hundred thousand. There are about two hundred thousand people in federal and state prisons and a hundred and forty thousand in county jails at any given moment. But only forty-eight countries completed the census at all out of a hundred and thirty-five member states. In England, in spite of suffragettes, conscientious objectors and their modern successors, and in spite of a much more enlightened attitude on the part of the authorities, there is still room for more flexibility in allowing research into all aspects of prison conditions. In countries like Spain and Italy, let alone in South America, it is virtually impossible to find out what is really going on at all. And in spite of their talk and their showplaces the socialist countries are the most secretive of all.

At most we get occasional glimpses of what is going on from fragments of research, protests too spectacular to be hushed up, or from the accounts of former prisoners. What is certain is that the modern pessimism about penal institutions, like the optimism of the nineteen-thirties, can be found all over the world.

The French, at a time when wartime memories of prisons and prison camps were still very fresh in their minds, proclaimed a new deal for their own prisoners. A primary duty of their *Juges de l'Application des Peines* was to supervise what went on in penal institutions. Yet now there is not only open revolt in the prisons but a murmuring in the most responsible quarters about a penal crisis due partly to great poverty of resources but even more to the loss of the will and spirit which animated the reformers of 1945. Mr. Ramsey Clark, a former Attorney General of the United States under President Johnson, is sunk in deep and dogmatic gloom. Claiming that more than half of those sent to prisons are returned there, he christens them "The factories of crime". The American Quakers, once devoted to the cause of using prisons to bring the criminal to repentance, have flown to the opposite extreme, asserting that all they can or should do is to punish. Even Nordic countries, which have so long prided themselves on leading the way in enlightened methods of reforming offenders, have now reverted to the same idea. Prisons stand condemned both for their failures to rehabilitate or to deter and for their inhumanities and injustices.

Critics of prisons there have always been, since the day when Charles Dickens visited the model penitentiary set up by the Quakers in Pennsylvania and reported that it was "wonderfully kept, but a most dreadful, fearful place". But now there is a new unanimity. Everybody is joining in: journalists and politicians, academics, officials, even ministers and administrators.

As long ago as 1925 Sir Arthur Waller, then Chairman of the

Prison Commissioners for England and Wales, suggested to the
International Penal and Penitentiary Congress that a set of general rules should be drawn up, governing the treatment of prisoners in all the member countries. As often happens he, as the
proposer of the idea, was given the job of drafting the Rules,
with the help of two of his fellow Commissioners.

That was the beginning of what have come to be known as the
Standard Minimum Rules for the Treatment of Prisoners. Responsibility for them was taken over by the League of Nations in
the nineteen-thirties and by the United Nations after the war.
Though they have since grown in number and expanded in scope
their original aims sounded modest enough. It was recognised
that different nations were at very different stages of civilisation
and order, and possessed of very different resources. It was recognised that there was still disagreement as to whether the basic
purpose of imprisonment was expiation or reformation. So what
was sought was no ideal system, no splendid castle in the air, but
a practical goal, selected after cool consideration of the realities.
The rules were to represent a minimum, but it was hoped a
minimum capable of being reached by any country.

Half a century later, at the United Nations Congress on the
Prevention of Crime and the Treatment of Offenders held in
Geneva in 1975, it was clear that scarcely a single country could
honestly claim to have fulfilled even these basic requirements.
Only sixty-two countries, again less than half the member nations, replied at all to an enquiry on this matter. And the Secretariat felt bound to question whether their replies reflected actual
practice or the opinions and wishes of the respondents. Even so,
only half of those replying claimed to have met the basic standards laid down regarding accommodation and decent living
conditions, many admitted tremendous difficulties in providing
adequately qualified staff, and only sixty per cent claimed fully to
comply with the guidelines on discipline and punishment, designed to protect prisoners from arbitrary treatment. Yet there is
little doubt that the countries offering answers were those that
had least cause for shame.

The general level of prisons throughout the world has always
been appallingly low. In many countries they have simply been
allowed to stagnate: as the pressures of crime have increased,
they have sunk to almost unbelievable depths of callousness and

squalor, even to systematic and deliberate torture. The perfunctory, summary, often brutal state of criminal justice in many
developing countries holds out little promise of genuinely civilised penal institutions.

Nor can we find much comfort if we seek an index of progress
amongst the more advanced. Take Western Europe. Scarcely any
country has solved the problems of prison work and remuner-
ation, none those of after-care. In France the system is outdated
and under almost unbearable pressure. In Italy it is frankly de-
moralised from top to bottom. In countries like Spain and Greece,
where the vicissitudes of politics have combined with the strains
of poverty, little can be expected. England is amongst those
failing to comply with the first of the "minimum rules," that
prisoners should sleep in separate cells or supervised dormitories.
Though her social conscience about prisons remains awake, she
can barely hold on to the advances she has achieved. West
Germany is still having a hard struggle to transform the spirit, as
well as the machinery, of her penal administration. Holland and
Belgium are perhaps coming nearest to the standards laid down by
the comity of nations, but even they fall short in certain important respects. The penitentiary nightmare in the United States
has been described so often and from so many angles that no more
need be said of it. Local jails everywhere are scarcely out of the
nineteenth century.

It might have been hoped that the newer ideologies, the contempt for tradition boasted by the socialist countries, would have
invaded their prisons and brought something fresh into such
symbols of old-style oppression. But I have seen no evidence of it.
The disjointed scraps of information emerging from China give
little confidence. The latest account of the fate of prisoners there
is disturbing. It portrays an intensive version of the mass psychological pressures to conform that seem to dominate Chinese life.

The Changed Climate

Imprisonment has been with us for a long while. It has been a
common penalty for at least two hundred years, and for the last
century has been the standard ultimate penalty for anything short

of murder and high treason. Why is it only now that antagonism and criticism has become so widespread? Why is it that the old optimism, the old belief that it was only some remediable fault in the theory or the management that was wrong, has suddenly failed us? In the past, the failure of one system merely spurred reformers to try a new one. If herding prisoners together did not work, what was needed was segregation. If segregation did not work, what was needed was controlled association. If useless toil on the treadwheel did not deter, useful work in workshops would rehabilitate. If no impact was made by regarding prisoners as lazy or wicked, we might get somewhere by treating them as psychologically sick. Failing that perhaps the answer was education, or vocational training, or socialisation by way of group therapy. But today it is not only the way prisons should be administered that is called into question but the very institution of imprisonment itself.

This mood has complex origins and can be seen as part of a change in the whole social climate, in many parts of the world. The Nazi concentration camps, and the revelations about them after the war was over, cut deep into the roots of complacency. They had, after all, carried to its logical conclusion the practice of consigning the unwanted to oblivion behind high walls, or just behind barbed wire. Prisons now share the fate of most establish social institutions. They are quoted as just another example of oppression, of people being sacrificed to the protection of property or the privileges of the powerful. Education as a means to conformity has gone out of fashion, and with it acceptance of the role of prisons in training men to obedience and orderly habits. Modern ideas of questioning, criticism, individual originality and initiative, do not fit in nearly as well with the smooth running of a penal institution, the control of difficult, even dangerous, men. Far-reaching achievements in the physical sciences, in industry and technology, have exacerbated impatience with the shortcomings of the social sciences, their inability to produce workable solutions to major human problems. Failures to alter the attitudes and behaviour of prisoners, repeatedly demonstrated by research into recidivism, coincide with this wider scepticism.

If the impact of penal institutions is dwindling it may be attributable, at least in part, to the sheer increase in crime. That has hit prisons in several ways. There are the direct effects of over-crowding in mutual contamination, in limiting rehabilitative efforts, in increasing concern with security. There is the growing range of alternatives open to the courts, which means that, at least in the more progressive countries, prison is increasingly re-served for habitual, professional or dangerous offenders. Kant said that you can make a man better only by using the remnant of good that is in him. In many of those who still go to prison it is hard to get hold of the good and still harder to build upon it. And when men leave penal institutions they return to a society where crime and the opportunities for crime are more prevalent than in the past. Whether he is idle or in employment, the ex-prisoner is surrounded by opportunities for petty theft or fraud.

Then there are the effects of unremitting publicity. Accounts of prison life, the more lurid the better, have sold well since the days of the Newgate ballads. We still have prisoners' reminis-cences, whether written by themselves or embellished by journal-ists. In addition we are now kept up to date, in press, radio and television, with accounts of prison conditions, not to mention the revolts and complaints of the more spectacular prisoners. In En-gland the perpetrators of the great train robbery, the members of the Kray gang and the moors murderers are seldom out of the news for long. There are major national scandals, like the dis-covery of the battered corpses of prisoners in a shallow grave on a prison farm in a remote corner of the United States, backed by evidence of beating, extortion and blackmail. There are the re-searches of international organisations, like Justice or Amnesty International or the United Nations, bringing to light similar scandals, often on a far wider scale and prolonged over years, in many parts of the world.

The modern rejection of censorship has produced a situation in which the findings of academic researchers or the reports of official investigations can include accounts of brutality in terms that not very long ago would have been classed as pornographic. For example, when I talked to prison administrators in the thirties and forties, I was assured that homosexual behaviour amongst prisoners was not a problem. Now you can read a report on assaults in prisons and prison vans prepared by a Public Prosecutor's Office in the United States, which admits that "sexual assaults are endemic" in some prison systems, and goes on to give not only statistics but explicit examples. A gruesome story unfolds itself: blackmail, beating-up, homosexual relations forced upon young men, scarcely more than boys, accompanied by violence from whole groups of older prisoners. That report comes from Philadelphia, but does not apply there alone. Much the same stories could be uncovered in the prisons of many countries were it not that they still keep the veil of secrecy closely drawn. All this contributes to scepticism about what can happen in prison, to a sense of despair about the whole operation—similar, on a smaller scale, to the feeling of defeat we suffer when continually confronted with stories of oppression, atrocity and war.

The fact that prisoners are joining the fashion of protest is hardly surprising. There are those outside, as well as inside, who feel it their duty to urge them to do so, on the ground that nothing is changed so long as reform is left to well-meaning liberals or well-intentioned bureaucrats. It is a curious fact that rebellion can occur not only when the underdog is being ruthlessly trampled underfoot but when some moves have been attempted to acknowledge his rights and better his lot. The loosening of rigid control, the gap between the goal and its fulfilment, the feeling that much more could be seized with a bit more self-assertion, has provoked violence in many kinds of situation.

The researches of criminologists have also played their part in shattering any illusion that all is as it should be in even the best-run penal systems. Analyses of the relationships between prisoners and prison officers and of the ways criminals in confinement influence each other has brought out the inherent conflicts, pressures and contradictions of life in prison.

Finally, there is the questioning of the right of the state, or indeed any other authority, to try to change people's attitudes, particularly under conditions of confinement and coercion. Some of this can be traced back to the imprisonment of political or religious, or other dissenters, or leaders of nationalist independence movements, in many parts of the world. Revulsion against so-called brain-washing techniques has also played its part. Not only is it argued that prisons are largely unsuccessful in changing the attitudes of prisoners, but that they have no business to attempt it.

The Fate of 'Reformative' Measures

What has become of attempts to make prison regimes reforma-
tive rather than merely deterrent? Since a prisoner has to do
time, cannot that time be used constructively to get him to face
and change the attitudes that lead him to commit crime? Why
should not the various aspects of his life in prison—work, leisure,
education, contacts with prison staff, contacts with family or
friends outside, contacts with other prisoners, the social environment of the institution itself, be deliberately directed towards
his reformation? Surely so much control over every aspect of life
should be turned to good account?

Work and vocational training

It has long been an attractive idea, to administrators and the
public at large, that prisoners should not be a burden on the
community. They should work to support themselves and their
families, to save something to help them on release, perhaps to
repay or compensate their victims. Into the bargain, they should
thereby learn habits of industry and skills they could use to earn
an honest living.

A prison which at once pays its own way and teaches its inmates to pay theirs has long seemed a most attractive proposition.
In the sixteenth century the ingenious Dutch established an institution which put its prisoners in the mood from the outset. To
teach men that one must work to live, they filled their quarters
with water and forced them to pump it out to avoid being
drowned. At the height of the industrial revolution some English
prisons were run as successful factories, with all the advantages of
forced labour and draconian punishments, as well as a share in the
profits, to keep their operatives at their tasks. In the United States
the fact that the prison could be self-supporting, and even make a
profit, was an important factor in the popularity of the "Auburn
system", whereby prisoners worked together all day though
segregated in separate cells at night. Yet there are formidable
obstacles to running modern prison industries on a basis that
either brings in profits or offers prisoners genuine experience and
preparation for work outside.

The chances that a prisoner will be employed in industrial work, even for a few hours a day, are no more than fifty-fifty in England, even lower in the United States. If he qualifies for an open institution or for outside working parties, whether in England or the States, the likelihood is that he will be employed in agriculture or forestry rather than in the kind of work available in the urban areas to which he is most likely to return. In a city prison he may well find himself employed in an old, overcrowded workshop, constructed to provide the prisoners with an occupation rather than anything that can be dignified by the name of necessary and productive work. More probably he will be employed, or rather underemployed, in cleaning out rooms and corridors or helping the cook prepare meals. In none of these pursuits are the tasks or the tempo very likely to capture his interest or prepare him for work outside.

Part of the blame for this, as for so many other things, can be laid at the door of old-fashioned prisons, deliberately built to cater for useless work in solitude. There is no space for the workrooms and the machinery needed for modern industry. Part of the blame lies with the multiplicity of functions and the continual turnover of people in the local prisons: how can you run an industry with workers who are constantly disappearing? Then, too, many of the prisoners themselves are barely employable by outside standards, in terms of temperament, skill and intelligence. Some, it is admitted, simply dislike work, even if done by others. Then there are the problems that originate from outside: the suspicion that, if prisoners are given good working conditions, they are going to be better off than the unemployed outside, even take the bread out of their mouths. Although the attitude of trade unions has been modified in some ways, there are still examples of refusal to accept trade and craft qualifications earned in prison. And industrialists continue to oppose what they fear will be unfair competition.

In America today I have seen many federal prisons which look like factory units and been impressed by their equipment and the tempo of their work. It is from them we have taken the pattern for Coldingley, the single English industrial prison so far established. But the federal prisons account for only a very small fraction of the penal institutions of America. Coldingley can take no more than a selected three hundred from the forty thousand or more who make up the prison population of England and Wales. Of course there are other places and other schemes that offer something approaching normal conditions of work. In several countries, including parts of the States, prisoners are employed in the building or reconstruction of prisons, on road building and other public works. But although more vigorous and purposeful employment can make imprisonment a less deadening experience, there is a dearth of evidence that it influences the likelihood of recidivism or the chances of regular work after release. Much the same can be said of the many schemes of vocational training operated in prisons and institutions for young offenders. They appear to have little effect on subsequent records either of employment or of crime.

To pay a prisoner a living wage and require him, in turn, to pay his way, also raises great difficulties when more closely examined. To mean anything, it would have to be related to the real value of his work. But the productivity of prisons is hampered by their inadequacies of accommodation and equipment. Nor can they be geared to productivity alone: considerations of security and justice get in the way. If one man is engaged on productive work and paid accordingly, what of those living alongside him who are not suitable for such employment? Where an incentive scheme is introduced, how is it possible to cope with the wide discrepancies between prisoners in skill, mental or physical ability, even in temperament, without producing a sense of injustice? How does a married man feel if his pay is swallowed up in supporting his family whilst his cell-mate is single and can keep it all? It is no answer that these differences and inequalities occur outside prison all the time. When it comes to men cooped up together during their sentences, inequalities of treatment can provoke a special bitterness.

Moreover the spending of money is as important as the earning of it. Personal possessions and personal luxuries are for many people an index of what a man is worth. In prison he is allowed scarcely any possessions: he cannot buy, or take pride in, his clothes, cell furniture, tools. He cannot, legitimately, gamble or drink, he can smoke very little. With prison wages at their present level what he can spend on his family, if he wants to, is pathetically small. He can save for his release, but he gets no realistic practice in budgeting. Ironically he is shielded from the shock of inflation which will hit him all the harder as soon as he steps outside.

Individual welfare and 'treatment'

More attention to the individual problems and needs of prisoners by psychiatrists, psychologists and social workers has been seen as another means to reform and rehabilitation. Specialists have been introduced into prisons to assess, to advise on allocation to different kinds of regime, training or treatment, to offer psycho-therapy and counselling to individuals or groups, to make contact with prisoners' families and to try to help them with personal and social problems. Measured in terms of reconvictions after release, the results have in general been discouraging, though both in England and in Scandinavia small controlled studies of regular casework with prisoners have recently offered more hope. But it is a rarity indeed in prison systems to find either such consistent individual attention or evidence of lower reconviction rates as a sequel.

The idea that each offender should be assigned, after careful study, to the kind of regime best suited to meet his needs, encounters a whole series of obstacles in practice. No country has anything like enough psychologists, psychiatrists or other specialists in its penal system to carry out the kind of 'diagnosis' or treatment envisaged. Even where an investigation is carried out, a specific recommendation made, there may be no available vacancy in an appropriate institution, or the institution may be so far from the offender's home that any advantage is counterbalanced by the disadvantages of virtually cutting him off from his family. Nor can it be assumed that his need for treatment will take priority in deciding where he will go. The first and overriding responsibility of prison systems is security. If, for example, the prisoner is considered likely to abscond, or to be dangerous if he does so, that will outweigh any argument that his chances of rehabilitation would be improved by sending him to an open prison or allowing his work-release. Even his usefulness in a prison industry may hinder his transfer to another institution.

The requirements of security, of order and equality within a prison likewise militate against attempts by psychiatrists or social workers to help him to sort out his personal problems, understand

and accept responsibility for his behaviour. Individual interviews cut across the prison routine of work, meals, cells, head-counting. This difficulty can be exacerbated by jealousy and suspicion, on the part of custodial staff, of better-paid 'specialists' whose assumptions and objectives are not the same as those of the prison.

The position of the specialist in prison is in any case an uneasy one. He may be seen by the rest of the staff as an interloper, at odds with the regime, insisting on special treatment for his 'patients', cutting across the demands of equality, even security. He may be viewed with suspicion by the prisoners, as a stooge of the establishment, or welcomed as someone who can be manipulated into expressing their special grievances or getting them special favours. He has to decide whether he can make any real impact by working with prisoners individually, or whether he would do better to spend his time trying to advise and support the custodial staff, in whose hands lies their day-to-day, hour-to-hour welfare. If he chooses the former course he may feel defeated by the attitudes, actions, decisions of the rest. If he chooses the latter, he comes up against the difficulty of bringing about any change in entrenched procedures and ways of seeing things, a difficulty aggravated by the continual movement and transfer of staff within and between penal institutions.

Lord Mountbatten, in his recommendations on prison security in England, commended the work of welfare officers in helping prisoners solve their problems and so reducing tension. But it is not always like that. The interview with the welfare officer, the group counselling session, may well bring forward problems, stir up fears or hatreds, which, if tranquillity is the major object, might better be left dormant. You do not usually encourage a man to search his heart or change his attitude by just letting him do his time quietly and keep out of trouble. An interview with a welfare officer may be a useful way of allowing a prisoner to blow off steam when he is worried, but it may have precisely the opposite effect, stirring up feelings and leading to an explosion.

In spite of his uneasy position, however, the specialist coming in from outside has a vital role in helping to break down the monolithic isolation that is still characteristic of the majority of modern prisons. Whether he is an industrial organiser or a trade instructor, a teacher, psychologist, psychiatrist or social worker, so long as he retains his associations with the world outside, brings in its ideas and assumptions, he does at least something to crack the hard shell of separation. Often his very difficulties indicate that he acts as an irritant, a challenge. There is a place for that in any institution that has so much control over all aspects of peoples' lives and is so little exposed to outside criticism. Moreover, the specialist who moves backwards and forwards between the all-embracing provision and regulation of the institution and the hurly-burly of life outside has something to offer that staff more fully absorbed in the prison system have not.

Limitations on letters and visits, censorship and supervision, exemplify a similar dilemma. It may be forcefully argued that prisoners need to be kept in touch with the world outside: in particular that preservation and improvement of links with their families offer the best hope that they will keep out of trouble on release. Yet access to people outside prison may threaten security, increase the danger of escapes or disturbances, even the continuance of crime. The prisoner who gets bad news from home, without being able to do anything about it, may become violent or depressed. And, quite simply, to allow more letters and visits is seen as requiring more staff to censor and supervise.

Conjugal visits have become a favourite hobby-horse amongst penal reformers. Why should sexual deprivation be included amongst the inevitable pains of imprisonment? Would not such visits from wives and girl friends solve the problem of sexual tension and homosexual behaviour amongst prisoners? I was Chairman of a Sub-Committee of the Advisory Council on the Penal System which examined this question very thoroughly in connection with the treatment of prisoners who have to be held for many years in conditions of maximum security. Obviously it is here that the problem is at its most acute. On one hand they are confined, and therefore deprived, for a very long time, on the other the risks of escape have to be most rigorously excluded. In all the talk of conjugal visits at least three things have to be kept in mind. Most prisoners are in for six months or less. Most are unmarried or separated from their wives. And for those who do not come into the top security categories, periodic 'home leave' offers a much better and more natural solution than conjugal visits in the unfamiliar and embarrassing atmosphere of a prison.

For maximum-security prisoners the best we could suggest was a small-scale experiment, whereby selected prisoners who had stable marriages and were serving fairly long sentences could spend a day or weekend with their families in some kind of family hostel outside the prison walls. The fact is that we simply do not know whether, even with well-adjusted families, this kind of thing can help maintain links and reduce tension or will only make the situation worse. Apart from that, thought must be given to the effect on a prison as a whole, with many men who are separated from their wives, if they ever had any, or who come from disturbed and disordered backgrounds, or perhaps are psychiatrically very abnormal. Much, too, depends on the attitude of governors and staff, and their fears about security. Advocates of conjugal visits often cite the example of Mexico and certain Latin American countries. I have visited an avant-garde prison of this kind, where every prisoner is said to be entitled to a weekly three-hour visit from wife or girl friend. I found it rife with complaints that these women were obliged to bestow their favours on the guards before being allowed access to the prisoners—a modern version of the 'droit de seigneur'.

In Sweden unsupervised visits by wives or girl friends, with the possibility of sexual intercourse, are the general rule in open institutions and are usually permitted also for long-term offenders in closed prisons. It has been reported, however, that the women feel embarrassed passing the guards when making such visits, and home leave is seen as preferable, wherever legally permissible and practicable.

A matter of more general concern is the whole relationship between prison officers and those in their custody. Why should not the custodial staff play a greater part in looking after the welfare of prisoners? They are in more continuous contact with them than social workers or psychiatrists, they know more of what is going on in the prison, can see how the men react to it and, if necessary, intervene quickly. Moreover, their attitudes and the ways they use their powers are of crucial importance. If they could become more involved in understanding, helping as well as controlling, would it not make prisons more human places?

At best prison officers are hampered by the very fact that they are custodians and directly responsible for discipline. When a prison is crowded and the emphasis on security is strong, they have little time for anything but locking and unlocking, counting, escorting, supervising the essentials of life. They, like the prisoners, are largely cut off from the world outside. There is little chance of making human contact with the prisoners or even of being seen as human beings themselves. Some are content with this, others are fighting for training and recognition as welfare workers. There have been repeated attempts to break down the barriers of anonymity, by assigning prison officers to particular wings, or to small groups of prisoners, and encouraging them to get to know the men and take an interest in their welfare. Yet again and again one is disappointed to hear that an experiment that seemed to be flourishing in a particular prison has faded out a few years later, perhaps because of a security scare, perhaps because of overcrowding, perhaps following a change of governor. Most discouraging of all is the reflection that in borstals, where the numbers have always been kept within bounds and where relations between boys and staff have always been stressed as a factor in reformation, so little success is now being achieved. It is perhaps not irrelevant that, where prisoners do say they have had help from someone on the staff of a penal institution, they are likely to be referring to the work supervisors or trade instructors, the people with whom their relationship is most nearly normal, most similar to what they would have with a foreman outside, least associated with the fact that they are prisoners.

Neither the obstacles nor the meagre evidence of success in reducing recidivism absolves us from responsibility. Measures to provide adequate work and leisure, to offer counselling or treatment, to improve relationships, can at least make life in prison less destructive. In putting people into custody we assume responsibility for their welfare in all these spheres.

Group influences

The influence of prisoners upon each other has traditionally been seen as a major source of crime. John Howard in Europe, the Quaker prison reformers in Pennsylvania, alike identified the unrestricted association of all kinds of offenders and suspects as amongst the worst features of imprisonment in the eighteenth century. Solitary confinement, separate cells, enforced silence, during work or leisure, even hooding to prevent prisoners recognising each other, have been amongst the devices tried to prevent mutual contamination. In more civilised terms there has been the separation of younger offenders from adults, of those in prison for the first time from the more experienced.

The old picture of prisons breeding crime by a kind of direct infection, the confirmed criminals passing on their attitudes, contacts and tricks of the trade to the newcomers, nevertheless remains potent. Prison populations everywhere are predominantly populations of recidivists. Those sentenced for the most serious offences are their established long-term residents, and they set the tone.

Interpretations of the ways such a situation affects prisoners have become more sophisticated in recent years. The 'society of captives', as it has been somewhat grandiloquently christened and described by criminologists, has its own heroes and stooges and villains, its own code and its own sanctions—like a boarding school or an army barracks, or for that matter a trade union. There is debate about whether the attitudes it adopts are brought in from outside, a criminal culture in microcosm, or whether they are built up by way of compensation and response to the humiliations and deprivations of imprisonment. There seems no reason why they should not owe something to both. They include a degree of distrust and hostility to authority, a measure of mutual support, an absolute ban on 'squealing', often a series of rackets, and sanctions ranging from ostracism and threats to outright violence.

The strength of such a code will vary with the way a prison is run. If prisoners are kept strictly apart or frequently moved around, it will be far less powerful than if they spend a good deal of time together, have freedom to talk, are not very strictly supervised and remain longer in one prison. The most elaborate and circumstantial accounts of prison codes and cultures derive from the United States. In Norway, where the prisoners come from far less criminal backgrounds and are far more segregated during their sentences, comparatively little of this has been found; even complaints to the authorities, though disapproved, can be made with little fear of retaliation. English prisons could be said to come between the two extremes. There is a difference between temporary adaptation to the fact of being a prisoner, and more permanent absorption of criminal attitudes and ways. However great the influence of the prisoners on each other, its impact is far from uniform. It varies with different kinds of people and at different stages of their confinement. In general, as in any other institution, it is strongest in the middle of their stay, when they are most soaked in the atmosphere, most dependent on each other, and the outside world seems most remote. For some at least it weakens as release draws near and they begin to look outwards again, returning to the attitudes with which they came in, whether persistently criminal or generally law-abiding. Some, especially the very inadequate, may shun their fellow-prisoners throughout their sentences, hiding away in their cells as much as possible. Others will be feared or rejected by the rest, perhaps because they have committed offences which are distasteful and despised like molesting children, perhaps because they are abnormally violent and troublesome. Some may indeed find prisons academies of crime. Others may simply be frightened and repelled by what they see and do all they can to keep clear of such associations after their release. So it would be a mistake to generalise. But what no one really knows is the most important thing of all: how far does a prisoner's involvement with other offenders in prison really have a lasting effect on his behaviour after he leaves?

If, however, the interaction of prisoners is as influential as has generally been assumed, why not turn it to good account? Instead of seeing it merely as a source of contamination or an impediment to attempts to 'treat' or change individuals, why not find ways of harnessing it to the purpose of reformation?

It has occasionally been suggested, by people with experience of wartime prison camps, that prisons should provide only the barest necessities and leave the prisoners to devise a tolerable life for themselves, with such social amenities and such social order as they can create by their own exertions. That would teach them responsibility. It may well have worked amongst groups of disciplined men, making the best of prolonged confinement together as they had learned to make the best of the army. But the mind boggles at the picture of the same thing happening in prison, amongst the weak and inadequate, the confidence tricksters and the violent, the unbalanced and the calculating, the experienced professional criminals who know all the ropes and the comparative newcomers who know none of them. In prison the worst tend to dominate, if only because they are more ruthless and stay longer.

Even far less ambitious attempts to encourage mutual responsibility, by giving prisoners a limited right of self-government, a right to discuss, to criticise and suggest, perhaps representative committees with the task of running lectures or sports activities, come up against this difficulty. It is so easy for prisoners to intimidate each other and manipulate such systems. And even when that danger is avoided there is a sense of unreality. The vital issues of liberty, discipline, resources, cannot be put in the control of prisoners. Responsibility that touches no more than the fringes of life soon palls: it may only increase frustration, a shadow of participation without the substance. It can be no real substitute, little real preparation, for the kinds of responsibility a man has to face the moment he steps outside the prison gate.

There were high hopes, especially in the nineteen-fifties and sixties, of the kinds of group discussion known variously as group therapy, group counselling and group interaction. Prisoners are encouraged, during such sessions, to exchange views and criticisms, express hostilities. It is not only a matter of blowing off steam or reducing tension within the institutions. There have been more ambitious hopes that the processes of mutual criticism would enable men to see more clearly the significance of their own behaviour, learn to accept more readily the demands of social living. But this process cannot get far unless it is voluntarily accepted. Where it is left entirely to prisoners to decide whether they will join in, the chances are that it will be those least in need of change who accept the opportunity. Others may well be deterred by the general prison suspicion of any 'treatment' initiated by the authorities. Attempts to counter this by making attendance compulsory or a virtual condition of parole may well be self-defeating in terms of genuine participation. And though group counselling, in its various forms, is still widely used, research has, in general, failed to confirm hopes that participation would make a significant impact on subsequent recidivism.

The "therapeutic community" is another idea borrowed from the treatment of mental patients and tried out in a few penal institutions. Ideally this means that barriers between staff and inmates are largely broken down. All participate in running things, deciding who shall be admitted, what rules shall be accepted, how problems shall be met or defaulters dealt with. Inevitably the possibility of achieving such a community in prison is hedged around with difficulties. There can be little real control of admissions or departures, the staff cannot divest themselves of their responsibility for security and discipline. Though there is a growing belief that the impact of prison on a man depends on the system as a whole, rather than on isolated attempts to educate, 'treat' or reform him, the idea of transforming penal institutions into therapeutic communities on the medical model remains a mirage.

"I regard as unfavourable to reformation the status of a prisoner throughout his whole career; the crushing of self-respect; the starving of all moral instinct he may possess; the absence of all opportunity to do or receive a kindness; the continual association with none but criminals; the forced labour and the denial of all liberty. I believe the true mode of reforming a man or restoring him to society is exactly the opposite to all these. But, of course, this is a mere idea. It is quite impracticable in prison. The unfavourable features I have mentioned are inseparable from prison life." Those were the words of an Englishman, Sir Godfrey Lushington, Under-Secretary of State at the Home Department at the end of the last century.[19] There have been strenuous efforts in many directions and in many countries to prove him wrong. But many would argue that they have simply proved him right.

Why Do We Still Send People to Prison?

Every other year the Institute of Criminology at Cambridge holds a Senior Course for people engaged in the practical tasks of dealing with criminals: judges, magistrates and their clerks, administrators, prison governors, senior officers of the police, probation and social services. One participant was recently heard to remark that, by the time they had finished discussing research about prisons, he had wondered why anyone was still being sent there at all. Yet, just afterwards, they had held a 'sentencing exercise', and had each been asked to say what sentence should be imposed upon certain offenders. And they had all found themselves deciding upon imprisonment as the only solution where the facts seemed to indicate that the convicted man was especially dangerous or had committed a deliberate and serious breach of trust.

People are not sent to prison primarily for their own good, or even in the hope that they will be cured of crime. Confinement is used as a measure of retribution, a symbol of condemnation, a vindication of the law. It is used as a warning and deterrent to others. It is used, above all, to protect other people, for a longer or shorter period, from an offender's depredations.

The basic function of a prison is to hold its prisoners, for as long as the law requires. Failing that, none of its other objectives can be achieved. An escape is the most blatant and humiliating challenge to the whole system of criminal justice. Yet it is always a possibility. The only place from which no one ever escapes is a cemetery. The hunger for freedom is always there: there have been instances of a man slipping away even in the last month of a long sentence. And though the casual escaper may not be a particular danger and is usually quickly recaptured, the violent or professional criminal who plans his getaway may be at large for some time, and is particularly likely to resort to violence.

It is no wonder, therefore, that security is always in the minds of those responsible for the custody of prisoners, that it casts its shadow over all activities, including those designed to reform, or just make life more tolerable. Yet there is a widespread idea that the trouble with prisons nowadays is that they are too soft, that 'rehabilitation' has been allowed to take over at the expense of both deterrence and security. No wonder prisoners get out of hand, goes the argument, no wonder they do not fear to go back, if they are allowed to associate with their cronies, have good meals, books, magazines, radio, television, little work, no responsibilities, even home leave before they go out. If prisons were a good deal grimmer, and the regime a good deal stiffer, it is contended, there would be fewer prisoners and less trouble with those there are.

Ironically, the very fact of rising crime and rising committals to penal institutions ensures that for most prisoners conditions remain quite sufficiently grim. More than half of all those sent to prison in England pass the whole of their sentence in the old general 'locals', where they are liable to spend eighteen hours a day locked up in their cells—longer at weekends or if officers are too busy with other duties to supervise them outside. Twelve thousand of them have to share cells with others. The Home Secretary, Mr. Roy Jenkins, said in the middle of 1975 that, if numbers rose to forty-two thousand, conditions in the system would approach the intolerable. The intolerable now seems around the corner, with the news that that number was reached in November, 1976. In the United States the Report on Corrections issued in 1973 by the National Advisory Commission on Criminal Justice Standards and Goals admitted that over half the adult prisoners in American state institutions are in maximum security prisons, geared to the fullest possible surveillance, with all other needs of the inmates coming a poor second. Some of these institutions are described by the same Commission as "dark, dingy, depressing, dungeons" or as containing "massive undifferentiated cell blocks, each caging as many as 500 men". And almost as many prisoners are to be found in local jails and workhouses as in state prisons: the majority of the jails are old, dirty, without space for work, treatment or visiting. Those in cities have been described as dangerously over-crowded. "Their chief characteristics are crippling idleness, anonymous brutality, human degradation and repression."

The situation is even worse than that in many other prisons in the world. When I was in South Africa a few years ago I was taken to see what they regarded as one of their best prison colonies. They showed me round the farms, with prisoners picking flowers, felling trees, all smiling in the sun. Then we went to see the animals, the cows and horses and pigs. I was particularly taken with the piglets: three hundred and forty-five of them basking there, each in his own commodious little stye. Then I asked to see the prisoners' quarters. Forty or fifty men were packed together in one big cell, their mattresses piled on top of each other, a half-naked army of misery and vice. They were certainly not likely to be reformed: the prison staff were almost afraid to go amongst them for fear of being overpowered. The same kind of thing can be found in the prisons of South America. In some countries allegations of systematic torture unfortunately carry conviction.

It could be argued that no active measures are needed to make conditions in prison grim enough to be deterrent. The sheer weight of numbers is enough to ensure that. Besides, even under the best conditions, it remains true that imprisonment is, in itself, a severe punishment. Many kinds of deprivation are inseparable from it: the loss of status, dignity and independence, the loss of liberty, choice, possessions and responsibility, deprivation of normal human relationships, including contact with people of the opposite sex, with wife, husband, children, friends. Blankness and frustration accumulates with the length of sentence. The most that the so-called 'comforts' of prison can do is offer some alleviation.

Nor is there any evidence that prison becomes a more effective deterrent when its regime is deliberately made more severe. That was tried in England in the middle of the nineteenth century. Prisoners passed the first year of their sentences working in solitude in their cells, spending hours a day in useless toil on the treadwheel. When they were taken to chapel or exercise, they were hooded to prevent them from seeing or communicating with each other. Floggings were commonplace. Food was kept to a monotonous minimum. Yet, far from decreasing, recidivism amongst prisoners it grew to such a point that at the end of the century the whole policy was condemned.

The idea that short sharp sentences might be more effective than either corporal punishment or longer but less severe confinement has been tried in England more recently than that. In 1948 we introduced detention centres for young offenders. The theme was a brisk regime, demanding unremitting activity from morning to night and strict obedience to orders. It was intended to be unpleasant as well as bracing and it was intended to be a shock. Yet those in charge of the centres have remarked that after all, no one can be expected to go on being unpleasant to people for weeks and months on end. Do we really want a prison system in which prison officers devote themselves to being simply repressive all the time, perhaps for years? Bearing in mind that nearly all prisoners will be free again within a few weeks or months, do we really want to have amongst us people who have been undermined or embittered by deliberate humiliation and unnecessary deprivations? Alternatively, if we succeed in keeping out of prison all but the most persistent or dangerous offenders, who must be detained for long periods in the cause of protecting society, do we also want to condemn them to live in conditions amounting to death by inches?

Release

Concern about what becomes of a prisoner after his release has taken two forms. One is related to public protection: does he require continued vigilance and supervision to prevent further offences? The other is related to his own needs and welfare: should he be given special help in meeting any problems, social and economic, likely to follow his incarceration?

There are at least six statutory systems under which prisoners may be released. Some countries favour one, some another, but they are not mutually exclusive and several may exist side by side. They can be seen as involving increasing degrees of administrative discretion and of intervention after release.

Simplest is the release of a prisoner on completion of his full sentence without any conditions attached. Next comes release, still without conditions, after a man has served, without serious misbehaviour, a fixed proportion of his sentence: what the English call "remission" and the Americans "good time." Third comes early release subject to the condition that the prisoner may be recalled if he commits a further offence during the unexpired portion of his sentence. That is the usual system on the Continent of Europe. Conditional release can, in its turn, be combined with surveillance as in the nineteenth-century system of "tickets of leave" in England which required ex-prisoners to report regularly to the police. Beyond that again lies the system, characteristic of many training and reformatory institutions for young offenders, in which the date of release is related to response to training in the institution and a set period of supervision after release is officially regarded as part of the process of re-education and reintegration. In this context help and guidance is seen as part of the task of the supervisors, and licences carry such conditions as that the offender shall lead an "honest and industrious life." If he fails to do so he may, at least in theory, be returned to the institution even without having committed a further offence. The sixth, and in some ways the most important and controversial of all these devices is the system of selective release on parole, especially for longer-term prisoners, first launched in the United States and now widely used in most English-speaking countries. This, too, is concerned, at least in theory, with rehabilitation as well as preventive surveillance and may, especially in the United States, carry a wide range of restrictions on liberty. Recall to prison is thus put very much at the discretion of the supervisor and the parole board.

Parole raises many issues pertinent to the basis on which prisoners are, and should be, released. Its introduction in the United States sprang from the desire to counteract the wide disparities between sentences imposed by different courts for similar offences and offenders. It was seen also as a means of reducing excessively long sentences, thus relieving the pressure on prisons without necessitating statutory changes in maximum sentences or direct interference with the courts. There were likewise hopes that it would provide an incentive to good behaviour and co-operation in treatment. And it was argued that men who might deteriorate in prolonged confinement would respond better to the chance of rehabilitation in the community.

Now doubts about the whole system are growing, above all in its country of origin. Even those who still favour it have to acknowledge its drawbacks. Far from counterbalancing unjustified disparities in sentencing linked with the personality or circumstances of the offender, it may even aggravate them, prejudices of judges being shared by parole boards. Far from reducing the length of sentences, it may increase them, courts making allowance for chances of parole that may never become a reality. Parole as an incentive to co-operation can become, in practice, a weapon of coercion, a pressure simply to superficial conformity.

The prisons, instead of being warmed by hope, can seethe with tension, uncertainty, a sense of injustice. No one has yet discovered how to identify some crucial point after which further confinement will destroy, release, rehabilitate. Even if such a point is found in a particular case, other considerations, such as the gravity of the offence, may often militate against release. Moreover, if selections for parole are made, as they inevitably will be, with an eye to the risks to the community, those selected will tend to be the people least likely to give further trouble, already well on the way to rehabilitation. The prisoners considered most dangerous, least likely to find support in the community, will be left to serve their full time, then released without supervision or special help.

The very processes whereby parole is granted or revoked have likewise come under attack. From the point of view of the prisoner, since they involve the granting or refusal of liberty, they are as crucial as trial and sentence in the court. Yet they carry few of the same safeguards. In England, for example, the applicant cannot appear before those who decide his fate, he is not represented by counsel, he cannot challenge reports made about him, he will not be told reasons for refusal. In the States he may see a member of the parole board, but often not the board as a whole, nor can he claim legal representation. The official attitude is that a prisoner has already been sentenced: parole, if granted at all, is a privilege, not a right. It is a matter of executive discretion, not legal argument. Yet in both countries there are those who reject such a stand as hypocritical, insisting that legal safeguards are essential whenever a person's liberty is at stake.

One way round some of these difficulties is the idea of a fixed period of parole supervision, say a year or two for all who have served a substantial period in a penal institution, regardless of whether they are released early or not. That is the basis of after-care for young men leaving borstals or detention centres in England, and is one of the solutions now being proposed in the United States. But it would mean an extension of compulsory supervision and after-care to offenders who are likely to be still more resistant, and that at a time when there is increasing doubt about the possibility, even the propriety, of trying to achieve effective help or control on a compulsory basis at all.

Now there is a ferment of controversy. A few would retain, even extend, the existing principles and machinery. Some would refine the methods of selection for parole, add legal safeguards and appeal in case of rejection. Some would say the decision to release should be taken by the courts rather than administrators or ad hoc bodies of experts. Others would restrict consideration for parole to a small nucleus of long-term prisoners whose release hinged upon the question of their dangerousness to the public, leaving the rest to serve fixed sentences. Yet others would sweep away the whole idea of compulsory supervision and after-care, at least for adult offenders. There is the straightforward view that all prisoners should simply serve their sentences and be released without any conditions. And there has been the suggestion in England that the period of statutory remission should be increased from a third to forty-five, or even fifty, per cent of the sentence. I would regard it as a serious criticism that the report on young offenders by the Advisory Council on the Penal System in England, published as recently as 1973, fails to reflect this ferment.

To build a full range of safeguards into existing systems of granting parole would be a very complex and difficult task. It would amount to setting up a whole parallel system of sentencing. It would require a great many people to run, would work very slowly, and would be difficult to apply to all those eligible. Experience of the working of the system in the United States would hardly encourage advocacy of its adoption in countries which have so far managed without it. It is true that a great deal of thought has been devoted to devising standard criteria for parole throughout the United States, in the preparation of the Model Penal Code and, more recently, by the United States Board of Parole Research Unit. These standards will, no doubt, prove helpful. But the formidable problems of applying them in individual cases will remain, with all the concomitants of uncertainty and grievance in the prisons. And when the criteria are examined it is hard to avoid the conclusion that they add little to what the courts could or should have taken into account in fixing the initial sentences.

On the other hand to sweep away the whole system of parole in countries where it has become deeply embedded in the administration of criminal justice could make intolerable the already critical burden on the prisons. Some of those who advocate such a course would go back to the starting point and try to reduce imprisonment by lowering statutory maxima and issuing exhortations to the courts. Nevertheless, this course too has its dangers. Even if it succeeded, which is doubtful, it could easily be reversed in response to a continued rise in crime or other pressures. There would then be no way left of remedying excessively long sentences or reducing the strain on the prisons. To restore, in such a case, the counterbalancing machinery of parole would be a very long and difficult task. Indeed, once relinquished, the system could well be gone for good. The choice of whether to retain this established safety valve in spite of its imperfections and contradictions is a question of political judgment, a balancing of one set of disadvantages against another.

Whatever the basis of release, whatever the arguments about the need or justification for keeping ex-prisoners under a period of surveillance in the public interest, there remains the problem of their own special needs and interests in the aftermath of confinement. The idea that they require special care at this stage is as old as prison itself. Such concern for prisoners has traditionally been part of the sphere of personal charity extended to the poor, the sick and the maimed. It still retains elements of that philanthropic, casual, episodic past.

In nineteenth-century England there was a strong conviction that the tasks of rehabilitation could and should be left wholly to voluntary initiatives. It was argued that the last thing a released prisoner wants is further dealings with officials of the system that has punished him. If anyone was to offer him help, hostels, work or good counsel, it must be other private citizens. Moreover, there should be no compulsion upon him to seek or accept such help.

The extension of state involvement in welfare of various kinds, together with the heyday of enthusiasm for reforming offenders, changed much of this. Such officials as probation and parole officers are held responsible for rehabilitation as well as for surveillance. In England the probation and after-care service now has this duty whether the ex-prisoners concerned are under compulsory supervision or come for help of their own accord. In some areas especially they have made substantial efforts to extend their resources, vary their approach, bring in volunteers, enlist community support, expand the use of hostels, lodgings, employment officers and ancillary staff. Inevitably there are conflicts of purpose and expectations. Most ex-prisoners, if they ask for help at all, want it in material terms. Most probation and parole officers see something more as necessary, are concerned to tackle personal and social problems, attempt to change attitudes in the hope of improving an ex-prisoner's prospects, preventing recidivism. The clash of expectations can be exacerbated if there are signs of further involvement in crime. Even where progress is achieved it can be slow and very demanding in terms of judgment, skill and patience.

Moreover the sheer physical size of the task of after-care is immense. How are the thousands and thousands of men released every year, especially those who are homeless or without jobs, to be re-absorbed into the economy and social services of a country? This is a question that has never been squarely faced. I myself became depressed and impatient about the pace of advance in after-care in England. I recognised that the Report of the Advisory Council in the year 1963 marked a step in the right direction, in that it recommended acceptance of state responsibility for the after-care of everyone the state had imprisoned. But I dissociated myself from it in a minority report, urging much more radical measures to meet that responsibility. I should be hard pressed to find a single country in the world today which has a system of after-care commensurate with the scale and complexity of the task. Now, under the heavy clouds of recession this endeavour, like so many other specialised efforts, is likely to be swamped.

Prisoners' Rights

It used to be taken for granted that prisoners had scarcely any rights at all except those expressly accorded them by law or conceded as a matter of grace. But the last few decades have seen a marked change of climate, especially in the United States. In 1973 the Report of the Task Force on Corrections, appointed by the National Advisory Commission on Criminal Justice Goals gave pride of place to a fifty-page analysis of the rights of offenders, especially those in confinement. Discussions and litigation about their status and grievances has been opening up a wide new branch of applied penology with repercussions in several parts of the world.

Sources of prisoners' grievances are not far to seek. They are to be found in many aspects of their lives. There are elemental physical deprivations and squalor. There are gratuitous invasions of human dignity, prison regulations which are vague, arbitrary, unnecessarily restrictive or humiliating. There are unjustifiable impediments to free exercise of religion. There are excessively rigid restrictions on communication with the outside world by way of letters and visits. There has been victimisation of particular prisoners or groups, sometimes outright brutality, by those in charge of them. There has been inadequate protection against violence or sexual attacks by other prisoners. There have been instances of extreme measures of medical or surgical intervention without free consent, of pressure to participate in possibly dangerous experiments. There have been complaints that prisoners are penalised for refusal to participate in milder 'therapeutic' measures and, conversely, that promised rehabilitative treatment has not been forthcoming. Drastic injustices may be inflicted in the use of segregation or control units. Few prisons can operate without them, but who must be consigned to them, subject to what kinds of regime, for how long, and under what guarantees constitute a series of crucial questions. In all my travels I have had the greatest difficulty in penetrating the secrecy with which these units are surrounded. Finally, there has been the growing unrest about the general uncertainty, the procedures and conditions, bound up with the system of parole.

In the United States a key role in focusing attention upon the status of prisoners and in establishing their rights has been played by certain of the courts, including the Supreme Court. It was a court that laid down, over thirty years ago, the maxim that "A prisoner retains all the rights of a citizen except those expressly, or by necessary implication, taken from him by law." From that time other courts have moved a long way from the old position, still held sacrosanct in England, that what goes on in prison is an administrative responsibility with which they should not interfere. They have accepted, and often endorsed, complaints from prisoners about a wide variety of wrongs and deprivations. In doing so they have inevitably also accepted responsibility for distinguishing between rights removed as a "necessary implication" of imprisonment and rights which should be respected, or restricted as little as possible even in prison. They have thus become involved not only in dealing with cases of deliberate abuse, but with the more delicate and difficult problems of balancing the rights of prisoners against the needs of prison discipline and order. Sometimes they come down on one side, sometimes on the other. But the general trend of their decisions has been against blanket restrictions on such matters as letters, visits, freedom of speech, and towards the idea that strict security is necessary only for the most dangerous prisoners.

the role of the Supreme Court in the United States are those of the their responsibilities not only for the trial of offenders but for preserving the basic human rights laid down by the constitution.

In England and other European countries courts see their responsibilities somewhat differently. The nearest equivalent to the role of the Supreme Court in the United States is that of the European Commission and Court of Human Rights, in upholding the rights and standards set out in the European Convention on Human Rights. In response to complaints, they can deal with any breach of these rights, whether it results from a misuse of discretion in an individual case, or from the broader regulations or statutory provisions lying behind the alleged abuse. If the Commission fails to reach an agreed solution with the government concerned, it can refer the controversy to the Court of Human Rights for a final decision. Unfortunately, the Commission can accept appeals from individuals only if their governments have conceded the right of direct complaint. And, as in the matter of criminal law, though governments may be willing enough to sign conventions or to lay down acceptable standards and regulations, very few nations have yet gone so far as to permit their citizens, above all their prisoners, to appeal to an international authority against their own decisions.

In England complaints by prisoners to prison governors, or even to boards of visitors or the prison department, can be compared to complaints by citizens to the police about the police. The system has been defended and condemned by similar criteria. There are the arguments that only the prison authorities can really judge the exigencies of prison conditions or the behaviour of prison staff, and that nothing must be done to undermine their authority. And there is the counter-argument that they are too deeply involved to be judges in their own cause, the more so since prisons are still largely closed communities, prisoners even more at the mercy of the administration than suspects outside. They cannot initiate legal proceedings, or obtain legal advice in doing so, without the consent of the Home Secretary, even where their complaint is against a prison officer for whom he is responsible. They may fear that they will be accused of making false and malicious allegations against an officer, or of persistent groundless complaints, either of which constitute an offence against discipline. They are not allowed to approach their members of Parliament unless they have first gone through all the procedures for complaint within the prison system. And it is only through their members of Parliament that they can appeal to the ombudsman. The United Kingdom is one of the few countries to allow direct appeals to the European Commission. But for that, too, all internal avenues of complaint must first be exhausted and that takes time. Only one case against the British Government has so far reached the Court, the Commission having upheld a challenge to the rule that a prisoner may not consult a solicitor or initiate legal action without consent of the Home Secretary.

Stirrings of concern about the status and rights of prisoners have been felt in several other parts of Europe. Recent controversies in France reveal much misgiving, especially amongst the younger judges, about the operation of the continental system of *juges de l'application des peines*. I made a study of this device nearly forty years ago, when it was first launched in Italy by the head of Mussolini's prison department. On paper it looked a remarkable manifesto of legality: judges were to be given the responsibility for overseeing the way sentences were carried out, controlling the vital decisions. But it was soon only too clear that in practice it counted for very little. The role of the judges in the execution of sentence got a new boost when the French adopted the system in 1945. It spread from France to many other countries. As in Italy, however, official accounts of it have been quite uncritical, and even a leading criminological textbook by Professor Jacques Léauté of the University of Paris gives no idea of the reality. It has taken a series of major prison riots to bring into the open scandals which genuinely independent supervision should long since have ended. Now there is the feeling that judicial responsibility is no more than window-dressing, that the judges themselves have become absorbed into the penal administration and that this is yet another internal system of control that covers up much more than it reveals.

In several Scandinavian countries the grievances of prisoners already come within the ambit of independent ombudsmen. But some people would argue that even this is not enough. There is a movement to go much further. Attempts have recently been made to establish the right of prisoners to form trade unions, entitled to hammer out their share in the running of the institutions across the table with representatives of ministries of justice and prison authorities. Important as experiments in Scandinavian countries undoubtedly are, it is no less important to remember that the social and penal context in which these experiments are launched differs in many fundamental ways from those of the United States and even of the large industrialised countries of Western Europe. The very size of the prisons, the ratio of staff to prisoners, the length of sentences, the prison conditions, the background of the prisoners and the nature of their criminality: these and many other differences should serve as a permanent warning not to indulge in comparisons which may prove too facile. Apart from the experiments in 'self-government' or therapeutic communities within prisons, the idea of prisoners getting together to thrash out collectively their complaints or requests has generally been rejected as too dangerous. The controversy in the United States about the rights of Black Muslim prisoners to hold religious meetings sprang partly from the feeling that their religion had political and racist elements which could stir up ferment within the prisons. In England there has been the unofficial prisoners' union known as PROP (Preservation of the Rights of Prisoners) with a measure of support from outside as well as inside the prisons, though without official recognition.

Prisoners' unions raise many of the same difficulties as attempts at self-government by prisoners. Yet the lack of regularised channels for complaints about general conditions, as distinct from individual grievances, can lead to anything from the roof-top summer-holiday antics of a handful of prisoners in the Isle of Wight to the grim explosions of violence and counter-violence in the prisons of France or America. They are sensational, they attract at least passing attention, whether to nagging injustices or to almost intolerable abuses, but they seldom produce constructive results. Their causes and objectives tend to be fluid and confused; they frequently end by emphasising issues quite different from those that launched them. And they may well prove counter-productive, leading to even tighter security measures, even less real concern for prisoners' welfare.

There is a lot to be said for extending the role of appeal courts to safeguarding the rights of prisoners. It is the courts, after all, who send people to prison, should they not accept a measure of responsibility for protecting them whilst they are there? If, into the bargain, they learn more of how penal institutions are run and what can go wrong in them, so much the better. The complaint brought before a judicial tribunal has to be specific, both sides have to stick to the point, there can be calm, impartial weighing up of the relevant considerations, and a precise resolution of the controversy, capable of being built into a consistent body of policies and of rights. In the United States, incidentally, the movement towards evolving independent review of complaints has contributed to legal education: the defence of prisoners' rights by students of law has given them 'clinical' experience of offenders and the penal system never before available to them.

Obviously, the more the courts show themselves willing to espouse the rights of prisoners, the more complaints they are likely to receive. The Chief Justice of the United States has expressed some concern lest they get more than they can cope with. This recalls the fears expressed about giving more scope to appeals against sentence. It has been suggested that a possible alternative to direct appeal to the courts would be to set up quasi-judicial review bodies. They would not be tied to particular institutions but could visit the federal and state prisons on circuit to hear complaints. Only if there were disagreement on their decision would a complaint be taken to an appeals court. Most important, if they are to avoid the pitfalls of the system of *juges de l'application des peines*, they would need to remain closely identified with the courts, and not become absorbed into the penal system.

In general, the movement towards recognising prisoners' rights, evolving independent review of their complaints, has got much further in theory than in practice. No more than a beginning has been made. Even in the United States not all jurisdictions concur in the view that the courts have a duty to remedy complaints. Few prisons anywhere in the world fully practise the principles enunciated.

Nevertheless the movement for prisoners' rights has had a wide role in opening peoples' eyes to the state of prisons. In the last analysis the rights of prisoners can no more be isolated from the whole standing and condition of penal institutions than human rights can be isolated from the whole economic and social state of the world. Many of the most basic grievances cannot be alleviated by court rulings or any other system of appeals or negotiations. For them the only remedy, in so far as we must still have prisons at all, lies in far-reaching reconstruction.

On the most radical view, the problem of prison is prison itself, the only worthwhile reform of prisons is to abolish them. There is a temptation to turn our backs upon our penal institutions as beyond redemption, concentrating all our energy and zeal on the search for alternatives. Certainly there is everything to be said for avoiding unnecessary imprisonment and certainly, in the majority of countries, there is still ample scope for its reduction.

Nevertheless, with the worldwide growth of violence and of deliberate organised crime, we must come to grips with the fact that we have many people for whom prison remains the only option, whether by way of penalty or to safeguard the rest of the community.

A retreat from high-flown claims that prisons are designed to improve or rehabilitate offenders may serve one good purpose, bringing us back to concern with issues of basic decency and humanity in the whole range of our penal institutions. It is too easy to concentrate resources on elaborate experiments in treatment whilst ignoring the poverty of basic necessities. Or to build show prisons for some promising or interesting or dangerous category of offenders, whilst leaving the vast majority of prisoners and staff to struggle on in neglected jails or other decaying institutions.

Retrenchment is not despair. It may represent a retreat from impossible hopes but it also represents an assessment of what has realistically been achieved and the direction of fruitful endeavour for the future. There have been luminous pages, as well as very dark ones, in the history of prisons. But human schemes and systems, in all spheres of life, fall short of what is hoped, sprout dangerous or disastrous side-effects, grow stale with time.

We cannot simply rely on our heritage, whether of buildings, of ideas, of leaders. We have to build for ourselves using our own new awareness, our own new tools. That applies to prisons and the use we make of them, as much as to anything else.

The problems of imprisonment cannot be dismissed as marginal. I would even go beyond the estimate put forward by Professor Lopez-Rey in 1970, that there are a million and a half people in prison throughout the world. Moreover it is a growing problem even in countries where quite strenuous efforts are being made to counteract it. Nearly two hundred and fifty thousand people were incarcerated in the federal and state prisons of the United States at the beginning of 1976, an increase of almost twenty-four thousand on the previous year. The expansion of crime, especially serious crime leading to longer sentences, is a major underlying factor. And nobody can really know the numbers of political prisoners of one kind or another, shunted off by administrative measures to camps and reservations in the remoter corners of the earth.

When John Howard's book on the *State of Prisons* appeared, Sir Samuel Romilly (who was to crusade for the reform of the criminal law of England) read it at once. He summed up his emotions in a letter to a friend dated May 22nd, 1781: "Howard's *State of Prisons* is not a book of great literary merit; but it has a merit infinitely superior; it is one of those works which have been rare in all ages of the world,—being written with the view only to the good of mankind. . . . What a singular journey! Not to admire the wonders of age and nature,—not to visit courts and ape their manners,—but to compare the misery of men in different countries, and to study the arts of mitigating the torments of mankind!"[20]

It is our duty to recognise that this "misery of men in different countries" is still with us.

CHAPTER

10

Experimenting with Alternatives

The Search for Alternatives

If we are to stop sending so many people to prison what are we to do with them instead? There is never any dearth of suggestions. Put them on remote islands and leave them to it. Make them compensate their victims. Make them work. Bring back the stocks or the pillory or flogging or the death penalty. The pastime of devising nasty or appropriate penalties for criminals did not begin with Gilbert's Mikado. That sober utilitarian Jeremy Bentham spend many anxious hours working out the details of what he called 'characteristical' punishments, such as balancing footpads on spikes, mounting highwaymen on sharp iron bars, half-roasting arsonists and half-poisoning poisoners. The urge to find cheap but spectacular penalties that fit the crime and terrify the onlooker is as old as mankind. We made them compensate their victims in money or goods or labour, we confiscated all their belongings, we got rid of them as outlaws, exiles or slaves, we transported them or put them to work in the galleys, we whipped them, tortured, mutilated, branded or killed them. There are still states that do such things now.

The business-like modern search for alternatives to prison has motives at once more mercenary, more humanitarian, and more strictly practical. Penal institutions are very expensive to construct and maintain and have rapidly been becoming more so.

Today, in the United States, each cell in a new prison costs, on average, about sixty thousand dollars to build. And the yearly cost of containing and maintaining a prisoner is between fifteen and sixteen thousand dollars. Figures for England and Wales throw a startling light on the way expenses have grown over the past quarter century. In 1950 expenditure on the prison service was six million pounds. In 1975 costs amounted to one hundred and eighty-five million. Inflation, of course, accounts for part of the difference (the six million in 1950 would be roughly equivalent to some twenty-six million now), but the increase in the bill is nevertheless formidable. We know that prisons do harm rather than good to prisoners and their families. We know that at least six out of ten recidivists sentenced to prison are in trouble again within five years. We know that, prison or no prison, crime has doubled in a decade. Feelings of futility, as well as humanity, have provoked an almost desperate hunt for alternatives, not only in England but in Europe, North America, Australia and New Zealand. We have become a little like travellers bringing home new dishes from all over the world. Each country solemnly sends out its delegates, with instructions to look at the penal systems of California or Sweden or England or New Zealand and to bring back an exotic selection of measures to add to the menu at home.

In the past ten years or so England has already virtually halved the proportion, though not of course the number, of offenders sent to prison. With twice as many convicted of indictable offences in the magistrates' courts as in the early sixties, the total number imprisoned has gone up very little. Virtually all the rest of the rise has been dealt with by orders for discharge and probation, the new suspended sentence and, above all, two to three times as many fines. Out of every ten people convicted of indictable offences in magistrates' courts, over five are now fined, two or three are put on probation or discharged, another one gets a suspended sentence, and only one is either sent to prison or passed on to a higher court with a view to prison or borstal.

The final report in 1967 of the National Commission on the causes and prevention of violence in the United States stated, in relation to the more serious 'index' crimes, that about a quarter of those convicted were confined in penal institutions, the balance being released under probation supervision. It is such familiar humdrum devices as these that have become the backbone of the system of alternatives to prison. We take them for granted as essential if we are to continue, as it seems we must, to live with a great deal of crime.

Into the bargain, the simpler measures like fine and discharge, now used in England for seven out of ten of the adults convicted of indictable offences, have been shown to be reasonably sufficient. A follow-up in London demonstrated that only three out of ten of those fined, four out of ten of those discharged, were reconvicted within five years. That compares with six out of ten of the offenders, albeit presumably presenting greater problems, who were either put on probation or sent to prison. In the United States a fine is frequently combined with probation, a device not available in England except in the form of separate measures for separate offences.

The question at issue now is whether we can press the case further. The prisons are still over-crowded. There are still many offenders whose crimes, whose dangerousness, whose persistence, whose inability or unwillingness to co-operate, seems to make imprisonment the only course. But is it? Probation, for example, apparently produces no worse results, in terms of offences, than prison, and, on the face of it, does far less harm all round and costs far less in money and resources.

Bearing in mind that we are now seeking alternatives for the residue of more difficult offenders, the people who arouse public resentment or alarm, what further measures, short of imprisonment, can we suggest? Can any of the old ones be strengthened or stretched? Are there new ones waiting to be tried? Is it just laziness, or what has been called the punitive obsession, that keeps us dragging on with the old accustomed penalties, or are there insuperable obstacles in the way of sweeping change?

Monetary Penalties

In societies as obsessed with money as modern developed countries, where nine out of every ten crimes are crimes against property, surely the neatest most economical penalty consists in taking money away. If we can go beyond that and put it towards repairing the damage done, to the victim or to society, so much the better. If fines are already used, relatively effectively, for virtually all minor offenders and half the more serious, why not stretch them to cover three-quarters, or more, even of those?

Even in simpler communities where the payment of compensation or a fine has been the staple penalty for offences, there seems, nevertheless, to be a select group of crimes stigmatised as 'boteless', in the Anglo-Saxon phrase: too atrocious to permit the offender to buy his own safety or to have it bought for him by his family or friends. The same feeling persists in the most developed societies. There is revulsion against the idea that a person who has wilfully inflicted serious physical injury on someone else should escape with no more than a payment in money, however large. We may have reluctantly abandoned the notion of an eye for an eye, or penalties like flogging, but deprivation of physical liberty is still seen by many people as an appropriate personal retribution, quite apart from its temporary efficacy in containing further violence. At least in the present state of public opinion, the fine can hardly be extended to cover serious physical violence.

Then there is the major, cold-blooded fraud, or the organised robbery where no actual violence takes place though it may be threatened. Are we prepared to accept a purely monetary penalty, however harsh as due recompense for these? The latest English Criminal Justice Act has introduced the concept of criminal bankruptcy as a means of extracting the cash from offenders who have profited from such crimes. But Lord Widgery, Chairman of the committee which recommended the measure, was not optimistic. He thought it would catch only a small fraction. And the Act does not propose to substitute bankruptcy for imprisonment: it envisages it as an additional penalty. Otherwise the criminal bankrupt might be in no worse a position than his honest but unfortunate colleague whose business had failed through ill luck.

Moving down from these major crimes, and looking at the whole range of offences, there are the far commoner cases of people who regard fines as little more than occupational risks, virtually licences to carry on with their illegal practices, a minor tax upon profits. Then, too, unless fines are closely related to the means of those ordered to pay them, they are patently unjust as between individuals, and the injustice increases with the scale of the fines. A hundred pounds is a mountainous burden to a poor man, a flea-bite to a rich one. Courts in England are required by law to have regard to means in imposing fines. To some extent they do so, but not nearly enough. In Sweden they have the system of 'day fines' whereby, in more serious cases, the amount to be paid is a proportion of a man's net income, allowing for his necessary expenses. The gravity of the offence is reflected in the number of days for which this sum has to be paid. That seems fairer, but is very cumbersome to work out, especially in complicated societies with their complicated structures of pay and taxation.

Then, too, there is the inevitable difficulty, exacerbated if we try to stretch fines to cover more serious offences and to that end make them harsher, that the better off will pay, whereas the worst off and the feckless, even with every allowance for their means, are still likely to default and so end up in prison. In England legislation to combat this has been building up over nearly a century. The magistrates' courts are required to allow payment by instalments. Before committing anyone to prison in default of payment, they must see him and hear what he has to say, and consider information about his circumstances. As alternatives to prison, they can place him under supervision if they think he needs help in managing his money and meeting his obligations, or order the payments to be deducted from his wages if he is in work. Such measures, together with more general prosperity, have drastically cut down, but have still not eliminated, the trail of people imprisoned either because they have not paid fines imposed, or because magistrates know they have no means and that fining is useless.

None of this alters the fact that the fine has proved applicable to a very wide range of offences and offenders. It is economical, both in money and manpower (though some harassed magistrates' clerks trying to extract payments from far-flung motorists might question that). It does the minimum of social damage to the offender and his family, since it need involve no interruption of his job, his leisure or his home ties.

If the fine is so useful, is not compensation to the victim an even better idea? Why is it so common and acceptable in simpler societies, so little used in the industrialised world? Surely if money is to be paid for an injury it should go towards compensating the person directly injured? And surely it would be better for the offender to feel that he had done something to recompense his victim?

A common complaint about penal reformers is that they are so concerned with the needs of the offender that they forget about his victim. But it was Margery Fry, an ardent penal reformer, who led the campaign, after the second world war, to bring compensation for victims of crime fully into the ambit of English criminal law. She thought in terms of criminals paying over money to go direct to their victims, with moral benefits, as she saw it, to both sides. There had long been limited provisions for this kind of thing, related to certain specific offences, or linked with discharge or probation. The Criminal Justice Act of 1972 greatly extended and generalised them. English courts can now order that the offender shall compensate his victim, whatever the offence or sentence and they are making increasing use of this power. Several states on the Continent of Europe have similar provisions: damages can be awarded against the offender as soon as he is convicted. Four Australian states have recently adopted the same device.

It is notable that in the past, this kind of provision was very little used by the courts, even where it existed. Perhaps they tended to feel it unfair both to punish an offender and to make him repay. Perhaps they clung to the fiction that the victim can easily apply for damages in a civil court. Or, alternatively, they may have felt it unjust to allow those who can and will repay to escape other punishment. These objections may still affect them. But the greatest stumbling-block is practical. Vast physical or material damage can be done by an offender who has little money and little prospect of ever earning much either in prison or outside. A hasty blow, a smashed vehicle, a moment of mischievous arson, may involve loss or injuries costing many thousands of pounds to put right. The prospects for the victim are not hopeful.

As a result, the major schemes for compensating victims worked out in the more developed countries have provided for payments by the state rather than the offender. New Zealand led the way in 1963, closely followed by England in 1964. Certain American states, first among them California, were not far behind, though their schemes were in various ways more limited.

Eight out of the ten Canadian provinces have followed suit. It has been an impressive movement. The English Criminal Injuries Compensation Board paid out awards totalling nearly three million pounds in 1972 to victims of violent crime, more than any other such scheme in the world. But the nearest approach even to a generalised link between payments made by criminals and restitution to their victims was the device, reported from Cuba before the Castro revolution, of putting fines, confiscated property and a proportion of prison earnings into an indemnification fund. And that was sometimes too low to meet the demands on it. So again we have, on the face of it, an excellent idea, but one which raises problems of fairness as between offenders and still greater problems of practicality.

Work and Leisure

If offenders cannot pay in money, why should they not be made to pay in time or in work? If work in penal institutions is so unsatisfactory, why not expect people to work outside prison, either for the benefit of the community at large or for that of their victims? In the bad old days and the bad old places, criminals could be sold into slavery, or sent abroad to work on plantations or in galleys, or conscripted into the army or navy, or made to work in chain-gangs on docks, roads and public buildings. Without going as far as that, is there not a core of solid sense about the idea? Not to mention a solid unregenerate satisfaction in the thought that those who have tried to get money without working for it should be obliged to toil like the rest of us.

The Russians and Chinese, with their vast countries and great areas of undeveloped land calling out for hard labour, are in a strong position to exploit work as a penal and 'correctional' measure. Their vast hinterlands offer opportunities rather like those that were available to Britain in America and Australia in the eighteenth and nineteenth centuries. Geographically, economically, ideologically, these vast socialist countries can regard work as a sovereign remedy, appropriate for political dissenters at least as much as for those who would be classed as criminals or wastrels by Western standards. A Soviet work-camp, however, can hardly be looked upon as anything but another kind of prison. And the Chinese device of sending off hooligans and other urban offenders, deviants or suspects to work in the fields with the peasants would hardly fit in with the complexities of modern agriculture in the West.

The United Nations Congress on the Prevention of Crime and the Treatment of Offenders held in London in 1960 brought in a curious scatter of reports on attempts to use work as a penalty in various parts of the world. The Italians suggested that offenders should be allowed to pay off fines by labour instead of imprisonment. The German penal code of 1920 had permitted this, and they had tried to apply it in forest areas where work was easily to hand. But even there the practical difficulties had proved too great. The Spanish, who already had a system of remitting part of longer prison sentences in exchange for work (at the rate of two days' work for one day's prison) were considering compulsory labour without imprisonment: but the idea looked less progressive when they admitted that the workers would have to spend their nights in detention. The Swiss already had a provision for work instead of imprisonment written into their penal code, but had never used it in practice. The Greeks and Yugoslavs were convinced that it could not be operated satisfactorily so long as the offenders were allowed to remain at large. Yugoslavia had really tried: following the Soviet example she had introduced 'collective labour' as a penalty but abandoned it in 1951 as unfair, unworkable and ideologically unsound.

In some other parts of the world they had had, or hoped for, better luck. The state of Uttar Pradesh in India, resenting the expense of keeping people in prison in the midst of poverty and unemployment, wanted compulsory labour for two classes of offender. Those who could not pay their fines should be employed in public works whilst living in the ordinary way, and those who might otherwise have been sentenced to prison should suffer instead a combination of hard labour and house arrest. Exactly how these measures were to be made acceptable in a country where even the honest could not get work was not made clear. South Africa put forward a system strictly for the non-whites. Black men with short prison sentences could ask, instead, to be assigned to work for farmers. They received a very small wage and were kept by their masters in food, clothing and lodging. Tanganyika had likewise a scheme for permitting short sentence men to choose manual work as an alternative, in this case for public authorities only, spending the nights at home or in unguarded camps. They claimed good success: in 1957, out of five thousand assigned to work in this way, only four hundred eventually had to go to prison. But they were selective in their use of the scheme. The courts could rule out offenders they thought unsuitable, so could local administrative bodies; appropriate work had to be made available and the men themselves had to be both willing and medically fit. In the United States experimental alternatives to penal institutions for young offenders have sometimes included community work or service as part of their programmes of training. For adults there has been 'work-release', but it has been the return to the institution at night that has constituted the penalty, rather than the work by day. Recent 'diversion' schemes have set rehabilitation for and through work in the forefront of their objectives, but they have been seen more as ways of taking offenders out of the criminal justice system than as correctional devices available to sentencers. Essentially work is envisaged as a benefit rather than a penalty.

The British attitude to labour as a penalty in its own right has long been highly ambivalent, and that in more ways than one. There was heated discussion in the eighteenth and nineteenth centuries about proposals to put convicts to hard labour in public places instead of consigning them to prison, the hangman, the hulks or Australia. The idea was rejected for two quite contrary reasons, each advanced with vehement indignation. First, it was unthinkable that freeborn Englishmen should be subjected to such an indignity. Second, it was outrageous that useful work should be made a penalty at a time when industrious habits were being extolled as the mark of the worthy citizen and when many worthy citizens were in poverty through lack of jobs. Nor was that the only paradox. Even whilst these points of principle were being debated, freeborn Englishmen were undergoing forced labour in dockyards and arsenals at home. And those transported to America and later Australia endured the whole gamut of devices, from assignment to work as virtual slaves for local settlers to chain gangs on government projects.

More modern concepts of work as an alternative penalty are to be found in the periodic detention, introduced in New Zealand in the nineteen-sixties and, still more, in the community service orders launched in England in 1973.

Periodic detention was originally reserved for offenders under twenty-one. It involved spending weekends, and also one night a week, at special centres. In addition to other educational, thera-peutic or disciplinary features, work of value to the community, often outside the centre, was an important part of the regime. Several of the centres subsequently introduced for adults have been non-residential, with the emphasis upon work, sometimes inside the centres, but usually outside in parks, hospitals or other public projects. The system has not been without its troubles, especially when it has meant that young offenders have been brought together at weekends. And, whatever the original inten-tion, it is not now regarded primarily as an alternative to prison in the case of adults, but simply as an addition to the measures available to the courts.

The English community service order, under which offenders can be required to spend up to two hundred and forty hours of their leisure in unpaid work for the community, owed some of its inspiration to the New Zealand example. But it carried further the concept of work as a penal measure in its own right and was hailed as a bold and original experiment in this respect. After two years' testing in six very varied areas it has been accepted as a workable device. During the experimental period courts showed some confidence by using it, the organisers demonstrated that they could find or stimulate local voluntary organisations with whom offenders could work, and there were no examples of misbehaviour by offenders during their assignments sufficiently strident to alienate courts or public opinion. Accordingly the measure was extended to most of the rest of the country in 1975. Its advocates claim many advantages for it. Community service orders could keep suitable offenders out of prison and give them a chance to make restitution to society. In addition they would benefit from the opportunity to regain their self-respect, would learn the satisfaction of service, might work alongside volunteers, perhaps catch their spirit and continue in voluntary service after the orders expired.

In the eyes of some people this diversity of objectives has been a major factor in securing acceptance of the measure. Courts or public who may wish to see community service as primarily punitive or restitutive can do so. Probation officer organisers who want to see it as primarily rehabilitative can likewise do so. Offenders can respond to it as placing obligations upon them which are specific and limited, even if demanding. That applies, whether they merely regard it as a penalty or as a measure that they find more comprehensible, constructive and helpful than probation. On the other hand, it is a sentence which brings out, very strongly, familiar disagreements about the penal system. In its earliest days one of its supporters could be writing that 'the intention is not to humiliate or even to punish, but to promote a sense of social responsibility', whilst another was protesting 'it is rather important that the public should be satisfied that people are not lightly being spared prison sentences.'

What has so far been achieved, in winning acceptance from courts, voluntary organisations, work supervisors, public and offenders, has owed much to the enthusiasm, resource, imagination and persistence of the handful of pioneers, and to willingness to commit resources to the new venture. Whether that standard and impetus can be maintained as the measure is more widely extended remains to be seen. Even in its experimental stages, though in some areas it was indeed used mainly as a genuine alternative to custody, in others it took its place, like the New Zealand periodic detention, simply as a form of penalty or treatment to be used whenever appropriate for particular offenders.

Now widespread unemployment calls into question the original concept of community service orders as requiring leisure-time work by people in full-time jobs. For the unemployed community service loses much of its penal character: it takes a working man a year to work through an order of two hundred and forty hours. But that can be achieved by the unemployed in a few weeks. Should the workless therefore be debarred from this kind of sentence? Or should they be required, as is happening in some areas, to work out their time in normal leisure hours? Or should they be encouraged to regard the whole thing as a rehabilitative exercise, perhaps preparing them for a return to regular work? Both in that sense and in relation to general financial stringency, the future prospects of this new measure are closely bound up with broader economic developments.

The line is very blurred between measures which emphasise

spare-time work, and measures which start from the idea of de-priving offenders of their leisure. That may be done either as a punishment or in the hope of teaching them better ways to fill it. As in community service and many other penal experiments, such orders have turned out in practice to incorporate both punitive and rehabilitative objectives. In England the original idea was rather like that of keeping children in after school as an alterna-tive to giving them the cane. Attendance centres, mostly run by the police, were set up, following the 1948 Criminal Justice Act, in many large towns and cities. Courts could send adolescents to them for a specified number of hours on Saturday afternoon. They all laid stress on punctuality, order, discipline and drill. But many also developed an interest in teaching crafts and hobbies, like a kind of compulsory youth club. Experiments in extending this idea to young men, using the time to teach them useful skills like decorating or first aid or motor-bike maintenance, have proved less successful and are likely to be dropped as the shiny new idea of community service orders has its day in the penal fashion parade. At the same time, with the ravages of football hooliganism before people's eyes, there are recurrent demands for punitive, even preventive, deprivation of leisure, like reporting at police stations during home matches. Rather more constructively, there are the suggestions that people guilty of driving offences should be compelled to attend for instruction during their spare time. In the United States attendance at driver education clinics has often been made a condition of discharge. West German police run courses for young traffic offenders. From time to time we get proposals that, as in some cases in the States, erring mo-torists should be required to work as night porters in casualty wards, so as to bring home to them the enormity of their heedlessness.

Leisure is still a highly valued commodity, a time for freedom. As an alternative to the complete deprivation of freedom implicit in imprisonment, the partial deprivation of leisure seems a good compromise, especially if it can be successfully combined with some kind of further education in the broadest sense. The prob-lems lie in finding the people and premises to carry it out and securing the co-operation of the offenders in attendance and learning. As with fines and all other alternatives, the problems become greater as we move into the more difficult offenders. We can chalk up some successes, especially with the young, but in an age increasingly resentful of discipline and authority it is not clear that we can get much further with rather older people of the kind we should like to keep out of prison.

Residence

Hostels go a step further, offering a place to live and work within the community, a base for assessment, for helping people to find a place in society. Long-term prisoners may be moved into them for the last few months of a sentence to ease transition to life outside. Particularly if they are homeless, they may need them after release, perhaps only for a few days or weeks, but perhaps for much longer. Courts may decide to make residence at a hostel part of a probation order. More recent has been the idea of hostel orders, independent of either prisons or probation, especially for those who need specialised help, such as alcoholics. And most recent of all have been the experiments with bail hostels for offenders who would otherwise have to await trial in prison.

The various kinds of hostel repeat themselves all over the world. Some of them have followed their traditional lines for many years. Others have branched out into intensive work and experiments like group discussion or group therapy. Yet others have gone further still, developing hostels as democratic communities with the residents working out rules and exercising mutual controls. But it remains true everywhere that hostels are available only for a tiny minority of offenders. The explanation is not to be found solely in resistance to newer, less punitive ideas in penal matters, or even to people's fears about having hostels as neighbours. In its nature any hostel that is to allow normal life, to be more like a home than a penal institution or an army barracks, has to be fairly small, has to have staff of good quality, costs a great deal to establish and to run. Those run on self-governing lines demand a certain unity of purpose if they are to succeed.

Then there is the point of view of the offenders. Especially amongst the homeless, the alcoholics, those most unable to cope with life outside, there is a fear of the demands of living within a close group. It is just those most in need of such experience and of intensive help who are likeliest to shy away from it. There is a sense in which too much ambition to rehabilitate can be self-defeating. Especially in England, where considerable experimentation has been going on in the past two decades, there is increasing recognition that hostels alone cannot meet the needs of the kinds of offender likely to find themselves in prison partly because they are homeless. For them a whole complex of facilities is needed, including bedsitters, communal houses, employment and education, sheltered workshops with social workers able to offer guidance and support. The need to seek out, establish and maintain such special facilities grows as general social provision shrinks in times of financial stringency. Yet the cost, in skill and devotion as well as material terms, is high. Such measures do not offer any cheap alternative to prison.

Status

Status is another commodity highly prized in competitive societies, as is public acceptance in communal ones. It is generally acknowledged that the fear of a court appearance, public conviction and a piece in the local paper is as powerful a deterrent as any threat of punishment for the average respectable citizen. The fear of losing his licence keeps the pub-owner or manager on his toes. In England you can be debarred from keeping an animal if you have been convicted of cruelty, or from having a licence to fish if you have offended against the Salmon and Fresh Water Fisheries Act. Disqualification from driving is a potent restraint in a society where ownership and use of a car stands high amongst most people's priorities, bound up with pleasure, prestige, convenience and often business as well. Professional bodies may strike offenders off their registers, depriving them of their right to work. People convicted of offences against the young may be prevented from obtaining or keeping employment in such occupations as teaching, or youth work. Deprivation of civil and political rights is a common penalty. In France offenders can be deprived of the right to vote, to be trustees, to join the forces, to inherit property or, quite frequently, to live in a specified area. In Italy and West Germany they may lose all political rights. In parts of the States they may lose many of the most prized civil rights, as husbands and parents, as possessors and managers of property or businesses. The totalitarian states with their control over all aspects of life, can go even further, withdrawing the right to work. Shame and public censure or admonition are powerful weapons at the lower end of the penal scale in China and Russia.

Disqualifications are often entailed in the very fact of conviction or imprisonment, regarded as logical side-effects rather than penalties in their own right. Yet their impact is such that we may well consider more thoroughly whether some of them might not suffice as alternatives rather than additions to imprisonment. In almost every instance they serve multiple purposes: those of punishment and deterrence but also that of prevention. The corrupt politician disqualified from office; the doctor or teacher guilty of sexual exploitation forbidden to practise or teach; the dangerous driver banned from driving; the violent husband prohibited from approaching or living near his wife; each of these is restrained in exactly the area where he is most dangerous to others, whilst retaining his liberty in other respects. Why use the clumsy and stifling blanket of imprisonment where comparatively minor restriction will do the trick and, into the bargain, can be maintained for much longer?

Of course it is not as simple as that. Disqualification from driving is a good example of the difficulty of enforcing such restrictions without close and complicated surveillance. Police in England frankly confess that a great many youths under disqualification get away with illegal driving, until there is an accident. Then the person injured is liable to suffer because the youth is uninsured, and so is consigned, almost, inevitably, to a penal institution, the very course it had been hoped to avoid.

Behind that, however, lies a broader problem. Like the ancient penalties of branding, mutilation or outlawry, disqualifications and loss of status, whatever their effectiveness in deterring others, can deprive the offender of his incentive to live within the law, even push him into further crime. Fully enforced, they can be oppressive.

Discharge, Suspension and Supervision

Each of the innovations that has been discussed has its evident dangers, its rather limited applicability. But what about the other major devices which, alongside the fine, have done so much to keep the prison population, until very recently, within manageable bounds? These are, broadly, binding over, discharge, suspension of sentence and probation. How do they stand now?

Binding over, the oldest and now least commonly used, involves requiring an undertaking that the offender will be of good behaviour, with the proviso that if he breaks his word he will sacrifice a named sum of money. It is a kind of moral and financial restraint, still sometimes employed as an alternative to the physical restraint of imprisonment, especially in cases involving damage, quarrelling or the threat of political disorder. Its value, of course, depends on the willingness of the culprit to accept the agreement, or recognizance as it is called. Many of the modern protesters, for whom it might otherwise seem very suitable, make a virtue of refusing it and insisting on prison.

Much wider has been the scope of discharge, with or without the condition that if a further offence is committed within a specified period the offender risks a double sentence, for his present offence as well as his new one. Discharge is, in general, a sensible and reasonable way of dealing with those whose crimes are not too serious and who are not expected to offend again. Those who object that people convicted of offenses should not just be let off by the courts should recall the dark figure of crime, the extent of police cautioning, the penalty and deterrence involved in the mere fact of conviction. There are justifiable doubts where, as has happened recently in the case of some children who persist in housebreaking or damaging property, discharge is used over and over again, and interpreted, quite accurately, as impotence on the part of the courts. In England this can occur where a juvenile court has already gone to the limit of its powers in committing a child to the care of a social services department, but that department has been unable to find accommodation in which the child can be kept under necessary control. There are similar situations, and similar concern, in many other parts of the world. But that reflects not upon the proper

practice of discharging offenders in appropriate cases but upon
the folly of legislating to keep people out of penal institutions
without also providing adequate alternatives for those who need a
measure of restraint, in their own interests and those of the public.

The suspended sentence means that the offender is sentenced to
a period in prison, or in some countries to a heavy fine, but
execution of the sentence is suspended, and eventually abandoned, if he commits no further offence within a prescribed
time. Its inventors and advocates on the continent of Europe
have claimed that it is characteristically the appropriate measure
for the free and rational offender. He is told quite clearly what
will happen to him if he gets into further trouble and it is up to
him to amend his ways accordingly. Initiated in Belgium and
France at the end of the last century, when probation was establishing itself in the United States and England, it has spread rapidly round the world as a cheap and efficient method of reducing
the pressure on the prisons, whilst still allowing the courts to
express forcible disapproval by pronouncing the sentence appropriate to the crime. Recently it has reached countries like
England, where it has been given a place alongside probation on
the assumption that there are certain offenders who will be better
deterred by a specific threat than by the vaguer warning of an
unspecified penalty embodied in probation and conditional discharge.

However that may be, the introduction of the suspended sentence in England had its teething troubles. In the first flush of
enthusiasm, magistrates' courts were required to suspend every
first sentence to imprisonment, except those imposed for violence
or upon people who had already served terms in prison or in
other penal establishments, such as borstal. They resented this
restriction of their powers to send people to prison as and when
they judged it necessary. There is little doubt that in many cases
they resorted to suspended sentence in circumstances where they
would previously have used probation or imposed a fine. And
where they would in any case have imprisoned, they made sentences stiffer to compensate for being obliged to suspend them.
When people thus dealt with offended again the courts had little
choice: in the absence of special circumstances, they had to put
the suspended sentence into effect. So it became clear that sus-

pension was, if anything, increasing imprisonment rather than cutting it down. Now the unpopular compulsion to suspend has been removed, but there is still reason to believe that the suspended sentence is being used as a substitute for probation, not simply for imprisonment.

Probation is a far more ambitious and adaptable idea than discharge or suspended sentence. In what may be called its original version, the court prescribes no sentence, but instead requires the offender to be under the supervision of a probation officer and maintain contact with him for a prescribed period. In England this may be from one to three years, in parts of the United States up to five years. Only if he fails to keep the requirements, or commits another offence, does he become liable to sentence for his original crime. Probation is essentially selective, designed not for those who can or will look after themselves but for those whose offences may be linked with the many kinds of personal and social problems that the psychologists, psycho-analysts and sociologists suggest are linked with criminality, and that so frequently crop up before courts and within penal institutions. The fine or suspended sentence may indeed be the best alternative for the man who knows and bears in mind the risk he takes if he indulges in further crime. Probation tries to offer a way out for the person whose life is in a bit of a mess, the person who might stop breaking the law if he could be induced and helped to work out a way of coping with life that would be more acceptable to other people as well as to himself.

Probation started precisely with the objective of keeping out of prison the rather pathetic, petty but persistent offenders, especially drunks, who still constitute one of the most obdurate problems facing penal reformers in search of genuine alternatives. In the course of its development, stretching over a century in the United States and England and now extended to many other countries, probation has sometimes been restricted to the young or to comparatively minor offences and offenders, some of whom have not really needed supervision or help at all. Now, on the contrary, with the introduction and expansion of simpler devices to meet the simpler cases on one side and the insistent pressure to keep more difficult offenders out of prison on the other, the movement is towards concentrating probation upon people who may be very likely to get into trouble without it, or even in spite of it. It has been suggested, for example, that it is wasteful to use it for first offenders, unless they show exceptional need for such help. It should be reserved for people who, by repeating their offences, demonstrate that they may be on the road to persistent criminality.

But is all this not wishful thinking? Even in England, where probation is as well staffed as anywhere, how can you expect an officer, with other duties to attend to and with something like fifty people under his supervision, seeing them perhaps once a week to start with, once a fortnight or less thereafter, to have time to get to know and influence more than a handful of them, or to make much real impact on their outlook and circumstances? Must not 'supervision', in the sense of knowing just what people are doing, keeping them out of trouble, be largely a fiction? What of the finding that, even in the nineteen-fifties when crime rates were lower, six out of ten London probationers were eventually in trouble again? What would happen if, as was recently suggested, probation officers were required to take responsibility, under 'supervision and control orders', for the sort of youth who at present goes to borstal or prison and who, in seven cases out of ten, commits further offences within two or three years of release?

Probation has already been stretched and reinforced in various ways to meet the needs of certain kinds of offender for whom supervision and individual help alone are clearly inadequate. One of its main virtues, indeed, is that many kinds of treatment can be attached to it by way of special conditions. In England it was conceded very early that a young probationer could be required, with his own consent, to live for a specified period in a probation hostel. Now that applies to adults. Then there is the possibility, again with the consent of the probationer, of requiring him to receive treatment for his mental condition, in or out of a mental hospital, for part of his period on probation. A new move has been the experiment with day-training centres, to be attended for sixty days of a probation period, designed to help the kind of offender who drifts in and out of work, cannot cope with his personal and family affairs, needs help with the basic skills of modern living.

Experiments have been carried out, both in the States and in England, to see whether more intensive contacts with probation-ers, more active intervention in their circumstances, will

strengthen the impact of probation, improve their chances o
keeping out of trouble. Where such measures have been applied
without differentiation, to offenders of varied personality and
attitudes, the general results have been disappointing, though
there has been occasional evidence of success in reducing recidivism amongst particular types, classified, for example, as 'amenable' to the kind of treatment, as being bright, verbal and anxious
Now, in England, we have the proposal that a new form of
supervision order, designed to keep more youngsters out of penal
institutions, should give probation officers, at least in the early
stages, much stronger control over where the offender lives and
what work he does, with the possibility of committing him to
three days' custody if he looks like getting into further trouble.

Probation officers in England, for all their enthusiasm for keeping people out of prison, have been viewing these latest suggestions for increasing coercion with some unease. Underlying this
is a far broader question about the whole idea of treating or
rehabilitating offenders, as distinct from simply punishing them,
leaving them to rehabilitate themselves or to accept help, if they
want it, from the general social services. Both the attempts to
impose treatment and control and the reaction against them have
been taken much further in the United States than in England, in
penal institutions, in mental hospitals and in probation. In England the tendency has been towards restricting the scope of
requirements to those considered necessary to prevent a further
offence. In the States conditions of probation or parole often
extend control into many areas of personal and domestic life,
imposing restrictions and demanding standards of compliance
that are beginning to be seen, not least by probation officers, as
unjustifiable invasions of individual liberty. There have been instances of officers refusing to enforce such conditions. The urge
to help, to treat, to reform other people has its dark side and its
oppressive consequences if not kept within bounds.

Another concern, perhaps peculiarly the product of present-
day questioning of the right of anyone, any group, to impose
its standards on others, is reflected in the complaint that, for all
their talk of encouraging people to choose for themselves, probation officers, along with other social workers, are trying to foist
middle-class values on the working-class people who make up the
bulk of those they supervise. One response to this criticism is to

stress that the probation officer must seek to discover what problems are important in the eyes of the offender and to achieve some kind of 'contract' with him as to how they will go about solving them. Another response is to argue that what are really needed are probation officers recruited from the same classes as most of their clients. Many of the latest crop of English probation officers, working-class or not, are as long-haired as their probationers, not much older and vividly aware of their social problems. In the States the majority are definitely short-haired, and carry guns into the bargain: some of them are as frankly hated as some of the police. The fact that probation officers, in England and in parts of the States, are becoming concerned not only with changes in their probationers but with necessary changes in social conditions and the system of criminal justice, is surely to the good. But it must be kept in mind that poorer people, working-class people, have as much at stake as anyone in the basic rules of criminal law. They are no more likely than any other class to sanction theft or serious violence. If anything they are harder on offenders than are the middle-classes. In that central field of crime the probation officer is not imposing alien standards. On the contrary, his probationers may well be shocked if he tries to be too liberal.

All the same, it has been demonstrated, both in the States and England, that the probationers most likely to benefit are those whose attitudes are not too remote from those of their supervisors, those who are able to accept the mutual confidence on which probation is, ideally, based. Obvious enough, but it leaves us with the old problem that just the people most in need of help, the people who cannot make relationships, have the most problems in all parts of their lives, are most likely to persist in crime, are the most difficult either to make real contact with or to help.

So there are experiments, on the lines of alcoholics anonymous and similar movements, to bring in ex-criminals to counsel current criminals. In the States this has been quite widely promoted under the title of 'new careers'. The idea is now being cautiously tried out in England. The double hope is that it can open up fresh opportunities to people willing to turn away from lawbreaking, at the same time providing offenders and others in trouble with counsellors who have first-hand understanding of crime, its pen-

alties and the ways out. It is easy to see the dangers and limitations. But at least it can hardly be said that we are not willing to
experiment.

Bail

By no means all those in prison have been sentenced to imprisonment. Some are awaiting trial and may be acquitted. Some,
though convicted, have been remanded for examination or further enquiries pending sentence, and may eventually be given a
non-custodial penalty. The numbers, the suffering and the damage involved are very substantial. Over the past ten years in
England and Wales, whereas recorded crime has risen by seventy
to eighty per cent, the rise in the number in prison under sentence at any one time has gone up by no more than seventeen per
cent. In contrast the number being held in custody before being
sentenced has risen by over a hundred and fifty per cent. And
though this can be attributed partly to longer delays in bringing
the accused to trial that is, in itself, a very disturbing fact. In 1974
the total remanded in custody was over fifty thousand: at any
one time they constituted almost fifteen percent of the prison
population. And that happened despite the fact that courts have
power to release a defendant on his own recognisances, with or
without sureties, with or without setting money bail, and in any
case without requiring the sum set for bail to be paid into court
unless actual default occurs.

In the United States it is worse. It has been said that half the
prisoners in the county jails are awaiting trial, and may wait for
weeks, months, years. Moreover, unless they have been charged
with the gravest crimes, they are there solely because of poverty.
They cannot raise bail, since that depends, in turn, on gaining
acceptance by a bail bonding firm and being able to pay its
charges. The problem of prisoners held on remand, in almost
every country of the world, often for very long periods, whilst
awaiting trial or sentence, has barely begun to be touched. There
they remain, victims of delayed justice, clogging the prisons, impeding all efforts to improve their conditions and regimes. An
international enquiry into the length of detention before trial is

long overdue, but to be of use it must be carried out in an honest way. During the past fifteen years, however, there has been a mounting campaign, both in the United States and in England, to develop alternatives to pretrial and presentence imprisonment. The best known of many successful experiments was the Vera Foundation's bail project in Manhattan, which demonstrated that, given adequate information about defendants, courts could safely release over half of those sent for trial, simply on their own recognisances without money bail. Less than one per cent failed to appear for the hearing. Similarly encouraging results have recently been achieved in London.

In dealing with homeless defendants, or those who have no strong ties with the community where they are charged, there may be greater fears that they will abscond. To meet this there has been exploration of the possibility of release subject to a measure of supervision, perhaps to residence at a bail hostel. If the court thinks they need medical or psychiatric examination defendants may be required to attend or enter a hospital, or to see a prison doctor, again as a condition of release on recognisance.

In the United States the Federal Bail Reform Act of 1966 authorised various such alternatives to money bail. In England, following the recommendation of a Home Office Working Party, magistrates' courts have been asked to seek full information about offenders before rejecting applications for bail. They have been urged to adopt a presumption in favour of release rather than custody. A bill at present before Parliament is designed to put this presumption into statutory form, at least in respect of unconvicted defendants. There are cases where remand in custody is inevitable. But unnecessary confinement pending trial can put a defendant at a grave disadvantage in terms of social stigma, loss of income and damage to family, reduction of opportunities to prepare a defence. This is one of the areas where resort to imprisonment could be drastically curtailed.

Diversion and Reinstatement

The humiliation of prosecution and conviction alone can cast a lasting shadow over the life of an offender, impeding his rehabilitation, a constant threat to his security in any subsequent employment or achievements. The dimensions of the problem are immense, but precise information is lacking. Estimates of the number of people in the United States who have some kind of criminal conviction range widely: from twenty to fifty million. In the poorest districts of the large cities, nine out of ten men and boys are believed to have been implicated. Before the recent Rehabilitation of Offenders Act, a million people in England and Wales, convicted of indictable offences ten years ago or more, and not reconvicted since, still lived under the threat of exposure. Even those who move to new districts, sever old ties, can never be sure their convictions will not be traced, perhaps brought up again after many years. The problems of living down an offence are increasing. It is not just the risk of a chance piece of gossip. As modern methods of storing and consulting records of all kinds become efficient, as the interchange of information between big agencies, public and private, becomes more common, the chances rise that a would-be employer or creditor will be able to trace a conviction, however long ago. The growing practice of requiring applicants for jobs or visas to complete questionnaires, including queries about criminal convictions adds to the difficulties. So does the extension of 'bonding' against theft by employees, insurance companies requiring to be satisfied that they have no criminal convictions. So does the fact that more and more employment is coming under the aegis of government departments, often more concerned to save themselves from criticism by excluding anyone with a criminal record than to set an example by offering openings to those who have served their sentences. The larger the proportion of convicted people in the population, the more widespread the handicaps.

In recent years there has been a corresponding upsurge of concern about this predicament. Especially in the United States there has been a proliferation of experiment in what the President's Commission on Crime referred to as 'diversionary devices': procedures and facilities for suspending criminal proceedings on the understanding that the offender accepts guidance or treatment from agencies outside the system of criminal justice. If he successfully carries out this undertaking, the proceedings can be dropped. For those at the other extreme, who have been through the mill of conviction and sentence, measures have been introduced to remove, after specified periods, any disqualifications entailed, to expunge old criminal records, or at least to prohibit all public access and public reference to them. In short, the idea is to enable offenders to live down their crimes, earn a clean sheet.

The movement to divert offenders systematically from the processes of criminal justice has produced a high tide of discussion and experiment in the United States. This has undoubtedly owed much to the revulsion against prisons and the delays and abuses of justice, especially in the lower courts. It has owed something, also, to the views of a number of criminologists that prosecution, conviction and sentence, the labelling of a person as criminal, increase the chances that he will see himself as such and therefore continue in crime. As trends of crime have mounted, and with them the costs of prosecution and of the penal system, there has been the seductive argument that it would be cheaper to divert as many offenders as possible, with little formality, to other services, some of them voluntary.

Hundreds of experimental programmes have been launched in the States. One of the most prominent and imitated has been the Manhattan Court Employment Project, started in 1968, which provided counselling, vocational guidance and job-finding services for unemployed men referred by prosecutors, on the basis that if they did well prosecutions would be dropped. There have been similar projects designed to take offenders after conviction: for example, adjournment of sentence for one to six months with referral to a community counselling agency, the reward for satisfactory progress being avoidance of a prison sentence.

Other countries, including Sweden and more recently Scotland and England, have gone further than the United States in diverting juveniles to child welfare bodies or other social services. Many American States have permitted diversion of drug addicts, before or after being charged and pleading guilty, to various forms of treatment or support, though often without evidence of lasting benefit. For chronic alcoholics, one of the best known experiments has been the Manhattan Bowery Project, again sponsored by the Vera Institute. Instead of arresting drunks, police patrols offer them a ride to the shelter, where they are dried out and given the option of further help. If it has not produced many permanent cures, the project has at least succeeded in gaining the support of all the official bodies involved and has reduced the number of arrests of down-and-out alcoholics in the area by eighty per cent. It too has been widely imitated, though one critic has observed that, in view of the huge numbers involved and the shortage of funds and resources for long-term rehabilitation, the best method of diversion might be for the police simply to abstain from any intervention at all except when there appeared danger of a more serious offence.

There has, indeed, been some measure of alarm over the uncritical enthusiasm with which such diversionary devices have been imitated in so many places. With a few honourable exceptions there has been no adequate research. We do not know whether the people dealt with in these ways do better in the long run than those passing through the courts or penal system. If they are in fact more successful, we do not know whether it is because only the most promising candidates are accepted for such cherished projects. We do not know whether people whose cases the courts would have dismissed, or punished only by some minor fine, are being coerced into accepting onerous conditions of treatment, even detention. Drug addicts, it is said, often prefer a short spell in prison to a long one at a treatment centre. Certainly many of these schemes lack legal safeguards. And the idea of referring people for rehabilitation before conviction, on the basis of suspension of prosecution is an uneasy one. Particularly when the power to continue proceedings in case of non-compliance is in the hands of police or prosecutors, it can develop a nasty cat-and-mouse flavour.

Finally, there is the question of the public interest. If diversionary measures become commonplace, is their availability likely to reduce general deterrence, increase the nuisance or danger to be apprehended from individual offenders? Much depends, of course, on the kind of people being dealt with. It probably makes little difference in the case of alcoholics or drug addicts. With children and adolescents widespread diversion may well lessen general deterrence, but to assess the eventual danger to be expected from the individuals concerned would call for a complex long-term calculation.

Nor can we accept without question the claims that these schemes relieve the strain on the criminal justice and penal systems by bringing in fresh resources of people and money. It may be asked whether they are not rather drawing off from the regular systems just the resources of initiative, imagination, skill and devotion they so desperately need. We want independent, voluntary experiments on a small scale. But it is a long job to evaluate them properly, to build up confidence, to demonstrate that they can go on working well when the first enthusiastic pioneers have disappeared and they have to be run by ordinary people as permanent arrangements, subject to all the boring old problems of established institutions. To transform them, within a few years of launching, into a whole system designed to by-pass the one we have is rather like building shining new suburbs for some people and leaving the rest to rot in neglected city centres.

With all these reservations, however, the strengths of the movement deserve recognition. There is no doubt that many offenders in the more prosperous classes have always escaped prosecution because friends or relatives or financial resources have been at hand to help them to make a new start, keep out of further trouble. Why should the poor not have a similar chance of avoiding the stigma that can so much reduce the chances of rehabilitation? One of the effects of recognising the extent of hidden crime should be to make us more merciful towards some of those caught up in the machinery of justice.

In most European countries convictions are centrally recorded, often as part of a citizen's social dossier, and the various disqualifications attached to serious crime could last for life. This is normally balanced, however, by provision that a conviction shall be expunged after a certain number of years, usually related to the gravity of the offence. Carefully considered recommendations to achieve a similar result were made in a recent report by the Canadian Committee on Corrections, under the wise chairmanship of Judge Roger Ouimet. It proposed that official criminal records should be available only to organisations requiring them for court, police or correctional purposes. Records resulting from summary conviction should automatically be annulled and sealed after a period of two years without crime following the end of sentence. Records of indictable offences could be annulled at the discretion of the National Parole Board on application by an offender who had kept clear of crime for five years following his sentence. The police should have access to it only if they could satisfy the appropriate minister that the case was of enough public importance to justify the concession. In 1970 the Canadian Criminal Records Act adopted the idea of marking the rehabilitation by a 'pardon'.

In England a combined committee, set up by Justice, the Howard League and the National Association for the Care and Resettlement of Offenders, under the chairmanship of Lord Gardiner, rejected the idea that records should be expunged or sealed, on the ground that access to them was essential to police, sentencers, researchers. But it made recommendations which, with some amendments, became the basis of the Rehabilitation of Offenders Act, 1974. Periods of imprisonment between six months and two and a half years are to be followed by a rehabilitation period of ten years; imprisonment for not more than six months by seven years; fines or community service orders by five years; borstal by seven years. In probation or conditional discharge rehabilitation is completed with the successful completion of the order. Subject to exceptions related to criminal proceedings and to certain occupations involving trust and the care of others, rehabilitation is automatic at the end of the prescribed period: the offender does not have to reopen the matter by making application or attending a hearing.

Whilst the benefits to the offender who has genuinely turned over a new leaf, are indisputable, there are inevitable difficulties. Authors and journalists have seen the new rules as putting the freedom of the press in jeopardy. Reference to an old offence once rehabilitation is completed can no longer be defended simply on the basis that the facts are correct: it will be necessary for a writer to prove, in addition, that he had no malicious intention in publishing them. Then it can be argued that electors or selectors of people for important jobs are entitled to know and take into account all information about a candidate. It is generally conceded that this is true where the offence was directly relevant to the work or the office being sought. Some frauds or corrupt practices may justify exclusion from certain positions of trust in business or public life. Violence or sexual offences against the young may justify exclusion from such professions as teaching, child care, youth leadership. But beyond these exceptional instances it is commonsense not to make rehabilitation more precarious by retaining lasting restrictions on the offender.

The Need to Temper Enthusiasm with Realism

What is the effect of all this range of alternatives and experiments? It is hard to suppress a certain fear that our enthusiasm may carry us too far. The basic choices before the courts are limited: the choice between confinement and liberty; the choice between a penalty, whether to be paid in money or work, and discharge with or without conditions; the choice between leaving the offender to make his own way or providing some form of supervision or treatment. Too many options may cloud the issues. Pulling the other way are the reflections that offenders, like the rest of us, are very varied, that each of the measures is likely to find a response amongst some of them. Then, too, if prison is so wasteful, must we not be thankful for measures that lengthen the penal scale, delay the point at which imprisonment is seen as inevitable? Even if only three or four out of ten abstain from further crime after some alternative form of treatment, they represent that many fewer to go on to a penal institution. That makes it worth trying, not just to relieve pressure on prisons but because we are dealing with human lives, the lives of the offender and often his family as well.

In the past there has been excessive optimism about prisons, not merely as deterrents but as institutions for treatment and reformation. In our disillusion about that, let us not pin exaggerated hopes on alternatives. All sorts of estimates are being bandied about of the proportion of prisoners who could be otherwise dealt with. In England there have been instances in which high court judges and prison governors have gone so far as to suggest that seven out of ten men in prison should not be there. More moderately, the National Association of Probation Officers has estimated that three or four out of ten would respond instead to more intensive forms of supervision in the open. A spokesman of the Howard League has put the figure at more than half. And now the National Association for the Care and Resettlement of Offenders has claimed that numbers could be reduced by a fifth in the case of men, a third in the case of women. I asked Professor Drapkin, head of the Criminological Institute in Jerusalem, what would be his estimate for Israel and the countries in that part of the world. After some thought, he put it at eight out of ten. When I was asked myself how far I thought this process of reduction could be taken in the United States, I hazarded a much more modest figure, though still at least three out of ten. But that, of course was no more than a guess.

It is of little value to try to give figures of how far imprisonment could be reduced in the world as a whole. Nor should facile conclusions be drawn from comparisons between the proportions confined in one country and another. In Western Europe it was calculated, a few years ago, that the Netherlands had less than twenty-three of its people in prison for every hundred thousand, whereas England and Denmark had over seventy, Germany over eighty. But to learn much from such figures we should need far more detailed surveys than we have of their sentencing structures, their institutional alternatives and the kinds of people sent to prison.

Some of the Scandinavian countries may be able to produce very low figures, but their social conditions, their levels and types of crime, are very different from those in the cities of the United States, of Germany or of England, let alone of South America, Africa or Asia. The Icelanders, asked by a foreign delegation to allow a visit to their prison, had to admit they had only one prisoner, and he was out at the time, playing in the band at a local dance. Iceland has about the same population as Southampton. Can we imagine that city having, at a given moment, only one of its citizens under sentence of imprisonment?

Countries vary not only in their crime but in their level of penal development and the possibilities of transforming it, in the quality of social services, the legal and philanthropic traditions, the human, educational and material resources on which they can draw. Many of the developing countries, for example, have long traditions of control by local communities. Before constructing massive penal institutions on the western pattern they need to consider how far they really need them. It is true that prisons are, in a sense, much easier to build than genuine alternatives. You can run up a prison in two or three years. And once you have the prison it is easier to send offenders there than worry about alternatives. It takes far more, in research, education, the slow development of sound traditions, to build up a well controlled system

of probation or even suspended sentence. A prison, in the last
resort, is a crude instrument of physical control. Devising alternatives calls for much more flexibility, variety, tolerance.

In the countries that already have substantial prison systems the
position is different, but still far from uniform. England is already
very advanced both in the range of alternatives at the disposal of
her courts and the extent to which they are used. I would agree
that fifteen or twenty per cent of those now sentenced to imprisonment there might still be dealt with outside. But even that
cannot be done quickly or easily. Building up the supply of social
workers, social services, of hostels, lodgings, openings for employment, of training and treatment facilities, is a slow and
laborious process. It has been found so even in times of prosperity and in fields other than that of criminal justice. Financial
stringency threatens a further brake. We still lack, for example,
the widespread hostels for alcoholics which the Criminal Justice
Act of 1967 anticipated might replace imprisonment for drunks.
We have, as yet, only a handful of bail hostels.

In many parts of the United States there is room for a much
more vigorous drive to extend alternatives. In France, where
probation is still in the early stages of development, even more in
Italy, where it has yet to emerge from the realm of planning to
that of reality, there is at least as much scope for expansion. In
Germany they have shown a little more enterprise but there is
still a long way to go.

Whenever the attempt is being made to build alternatives to
imprisonment into systems of criminal justice, several things have
to be remembered. Attitudes cannot be changed by ministerial
decree, by administrative exhortations or the resolutions of
philanthropic and learned societies. In some countries, such as
England, it is not more legislation that is needed but more resources. In many others, however, the transition has yet to be
made from ingrained legal traditions, over-restrictive penal systems. When a switch to alternatives is first made the effects are
immediately visible and unquestionably beneficial. But the further we try to carry the process the more difficult it becomes.
There is increasing resistance from the public, politicians, offenders themselves. Whereas there may well be a saving, in money
and resources, in keeping out of prison people who are relatively
harmless and capable of looking after themselves, the cost of

dealing with the more persistent or dangerous offender will always be high, whether it be met in prisons or by alternative devices. Hostels, for example, are expensive to build and maintain, difficult to staff. People suitable, trained and willing to devote themselves to this work are hard to come by anywhere. The necessary resources cannot be had on the cheap. Finally, there is a law of diminishing returns: the more persistent the offender, the more difficult the problems he presents, the less the chance that he will remain out of trouble. If alternatives are reserved for the more hopeful cases their rate of 'success' will be high: as they are extended to these more challenging people it will inevitably drop.

This is not to imply that we should be lukewarm about alternatives. When all has been said by way of warning or reservation, they remain the major hope of penal advance. We can, after all, take courage from history. Less than two centuries ago in England it was being affirmed that alternatives to the death penalty for theft would mean that nobody's property would be safe. Less than one century ago it was considered essential that children should go to prison before being allowed the privilege of a reform school. Half a century ago Sidney and Beatrice Webb, in the epilogue to their classic study of English prisons, concluded that 'so far as can be seen at present the most practical and the most hopeful of prison reforms is to keep people out of prison altogether . . . It has been discovered that we can keep people out of prison by the simple expedient of not sending them there'.

When we look at the misuses of prison throughout the world we are spurred to find other and better ways of dealing with lawbreakers. In this, as in other things, we need to temper our enthusiasm with realism, refuse to be deluded by the hope of cheap and easy answers. But we need to kindle the imagination and drive of reformers, rally the support of public opinion, behind a determination not only to find alternatives but to carry them through, test and sustain them as essential and viable parts of our social system.

CHAPTER

II

Some Implications

To live in a world of rising crime has both direct and indirect implications.

Even in the most law-abiding countries, we have to come to terms with the fact that we had tended to develop expectations of security for ourselves and our property which are no longer realistic. We have lost a measure of our freedom. We have been compelled to face the need for greater caution in our dealings with others. The possibility of falling a victim to crime, in its traditional sense, has increasingly to be taken into account in our daily lives. The extension of the political crimes of people against governments and of governments against people puts a much sharper edge on this, with its special overtones of ruthlessness, betrayal and unpredictability. And the white-collar crime of the powerful and prosperous, though it evokes less physical fear, erodes standards of economic life, inflicts severe financial losses, makes individuals, commercial undertakings, political parties and even governments an easy prey to sophisticated fraud and bribery.

Alongside these direct consequences of expanding crime there are the indirect.

The spread of traditional crime breeds public apathy and defeatism. The police fall behind in bringing offenders to justice, their temptations to cut the corners increase, dangers of corruption rise. The courts, endlessly bogged down by long waiting lists, are under pressure to hurry cases through, to accept devices which offer short cuts, and to neglect legal protections. Penal institutions, overcrowded with the untried as well as with the sentenced, see tensions mount and standards deteriorate. All those who have to deal with offenders are frustrated by the widening gap between what is expected of them and what they can hope to achieve.

The injection of political crime adds to the stress throughout the system. Even the liberal are impelled towards ruthlessness when faced with the ruthless manipulation of their institutions. From very many parts of the world, solid evidence has been forthcoming that torture is as commonplace a tool today as it was in the middle ages. Though directed primarily against opponents of governments it infects the whole apparatus for enforcing criminal law. It is ironic that, in 1975, two hundred years after Cesare Beccaria voiced the gathering protests against the use of torture by the despots of Europe, a United Nations World Congress on Crime should have felt it necessary to pass a resolution banning torture, and that they regarded this as their major achievement.

The impact of white-collar crime, or of growing public awareness of it, has been less brutal but equally insidious. Durkheim saw crime and its punishment as a potent means of reasserting and strengthening social values. The virtual impunity and the vast gains of these macro-criminals, on the contrary, has produced a widespread disgust and cynicism which cuts at the very roots of social values.

Today we have a much better understanding of the complex social and individual elements in crime. But we have no evidence to demonstrate that advances in prosperity, welfare and education have reduced its incidence or even restrained its rise. Nor has it been shown that more elaborate devices in the classification and treatment of offenders have made any impact upon the rates of recidivism.

In peaceful England, in the middle of the affluent sixties, Professor F.H. McClintock estimated, on the basis of trends then existing, that one man in every three would be convicted of an indictable crime at some time in his life. And that leaves out of account the very many offenders who escape conviction altogether. The wide-ranging survey of research prepared by Robert Martinson on behalf of New York State, and the recent report on the effectiveness of sentencing published by the Home Office Research Unit in England, have tended to reinforce feelings of pessimism about our ability to prevent relapse into crime amongst known offenders.[21] The latest analysis by the F.B.I. of offenders found guilty of serious crimes in the United States has revealed that at least half of them were already known recidivists and at least a quarter had already no less than five convictions behind them. These conclusions differ little from estimates of recidivism made in Europe from time to time over the past century. There seems to be remarkable stability about the proportions, regardless both of social conditions and penal practice.

The natural response to rising crime, declining standards of justice, failing hopes, is to hit out and demand change.

The tough authoritarian approach to the control of crime has its own ideological vitality. It aims to support a regime, to preserve a set of values, to hold a community together and maintain order within it. Yet it exacts a very high price of the whole society, as well as of those suspected or convicted of crime.

In all ages it has widely been taken for granted that crime must be kept down by sheer severity, if not outright terror. This attitude, even in its extreme form, is by no means dead. Some of us have seen it revived in the civilised heart of Europe during our lifetime. The Nazi regime in Germany regarded crime simply as expressing the antisocial will of an individual in revolt against the laws of his community. That will must be broken or annihilated by intimidation, retribution or elimination. The limits of what was counted criminal were at once vague and perpetually extending. Respect for individual rights was rejected as decadent, since the highest duty of everyone must be to efface himself before the State and its leader. As Professor Donnedieu de Vabres expressed it, "We see here the close correlation, the inevitable correspondence, of these two activities of government; its general policy and its penal policy, the first giving impulse and direction to the second." What was true of Nazi Germany was true also of Fascist Italy, though it was all on a much smaller scale: Hitler had his concentration camps for millions whereas Mussolini was content with one small island. But the inexorable attitude to those suspected of crime or political dissidence was the same under both regimes.

Today a ruthlessly punitive attitude has been gaining ground in many parts of the world. Some countries, though well on the threshold of development in other respects, remain in the dark ages in terms of crime and punishment. In others, terror and counter-terror, linked with political turmoil, have become the order of the day. In yet others there is a struggle to maintain a precarious stability by using the machinery of criminal law to hold down oppressed majorities. Then there are the monolithic regimes which justify relentless measures against dissidents as an essential stage in ultimate ideological fulfilment. There are yet others which have escaped the controls of colonialism only to impose new tyrannies. And there are countries where enforcement of the law, in all its stages, has been allowed to stagnate, become corrupt and oppressive, less through deliberate intent than through sheer poverty, weakness and neglect. At the turn of the century Enrico Ferri was cheerfully consigning the liberal principles of criminal law to the historical archives, alongside free trade as advocated by Adam Smith. He thought the battle to protect the individual against the state had been won once and for all. By now we ought to have learned better than that.

The 'get tough' attitude has been endemic even in free societies, flaring up strongly in response to rising crime and violence, though its more extreme manifestations are restrained by established liberal traditions. Its theme is, nevertheless, that the protection of the criminal against injustice should yield place to the protection of society against crime. The police, and others entrusted with enforcing the law, should be given a freer hand. There should be a tightening of the screw throughout the penal system; severer punishments, more strictly enforced and, indeed, a return to the death penalty. If only, in a common phrase, the kid gloves were removed and offenders were given their just deserts, they would be taught that crime did not pay and the rest of us could be saved from their depredations.

More sophisticated advocacy of the central role of punishment is being taken up by certain academics and policy-makers. Though they differ in many respects, they all start from the assumption that proportionate punishment is important for the protection of the offender, as well as for that of society. They would, accordingly, limit the discretion of courts in sentencing and of administrators in enforcement. In particular, they would abolish indeterminate sentences at the beginning of the process and parole at the end. They would also eliminate anything that smacked of pressure to accept treatment or to demonstrate reformation as a condition of release. In these things they may be called neo-classicists, expressing, in various ways, a return to the idea that the right and duty of the state, when acting through the criminal justice system, is limited to the imposition of punishment proportionate to the crime.

An influential group of American Quakers has protested against the use of the penal system, and especially penal institutions, to coerce offenders into accepting prolonged "rehabilitative" treatment. Their report, *Struggle for Justice*, sparked off a whole train of investigations, books, public statements and discussions. Certain leading Scandinavian criminologists have similarly called for a retreat from the idea that reformation should come within the scope of the criminal law and those who enforce it. In England there have been attacks, though milder than those in the States, upon the system of parole, with its elements of uncertainty and administrative discrimination.

Thus where, until very recently, the zeal to reform by adapting the sentence to offenders as well as to the offence was something taken for granted in progressive countries, now it is under attack both as being impracticable and as opening the way to oppression.

It cannot be denied that there have been serious distortions and exaggerations in sentencing in the United States. In too many cases there has been either excessive leniency or excessive severity. But, as a general rule, when it comes down to the practical business of sentencing, especially in relation to adult offenders and serious offences, the gravity of the crime and the protection of the public have been the decisive factors. To suggest that, because of unjustifiable departures from these principles, we need to exclude discretion from a substantial sector of our system, is irrational, retrogressive and inhumane. Whether we attempted to effect such a change by detailed prescription or by mandatory sentences, we should find, as history has shown again and again, that it would not work. In so far as it was enforced it would make for rigidity rather than justice. It would be defeated by the healthy reaction of the judiciary and, when its operation was seen in individual cases, even of public opinion. The remedy for distortions in sentencing does not lie in a cut and dried list of penalties. It depends upon measures to improve rationality and consistency in the way discretion is used and to ensure adequate redress when it goes astray. A whole range of devices and proposals to achieve this has been evolving in recent years. To quote a recent and important Australian committee on criminal law and penal methods, "It is too little realised that in a modern developed community it is impossible to have a simple sentencing system."

A period when criminal justice is subject to so much strain also breeds a mood of recrimination. This is only too understandable, in some senses inevitable. Stagnation and abuses are very real. But the criticisms of what exists are often combined with wishful thinking about what might be: dreams that we could police ourselves, abolish prisons or discover new approaches that would miraculously transform the social response to offenders. Moreover, recriminations, in so far as they are misdirected, are self-defeating. Generalised denigration is not only unjust but dangerous, more likely to aggravate than to improve. It is unjust because those dealing with crime and criminals are constantly beset by dilemmas which society as a whole has left unsolved, dependent upon resources that society has left inadequate. It is dangerous because people with the qualities and qualifications needed for these responsibilities are hard to find and to keep. A constant barrage of criticism and discouragement may well leave us with only rascals and bullies to undertake such onerous tasks. It is not only the suspects and prisoners who are entitled to their dignity and to respect for their rights.

Yet if the massive growth of crime of all kinds is allowed to continue, where will it end? Perhaps it is possible to offer a little historical reassurance. In Germany, after the second world war, everything seemed to be breaking down. In Berlin there were twelve times as many complaints of theft as there had been before the war, seven times as many burglaries and robberies. And that at a time when many people felt it useless to go to the police at all. In the dire shortage of necessities, money had lost its value and respect for property was shattered. Trains and lorries were cleared of their loads as they travelled, farmhouses, fields and barns raided by looting parties. Food, building materials, electrical equipment, disappeared wholesale. It was estimated that in 1946 crime had swollen to five times its normal scale. Yet in 1947, as the value of money began to be restored, as social and commercial normality returned, crime was already beginning to shrink. It was the bankruptcy of the state, the falling apart of society that

had released the flood of crime, not the flood of crime that destroyed the value of money or the authority of the state. Criminals do not determine what becomes of society: they merely
show an extraordinary power to adapt to the social climate,
whatever it may be. What is remarkable is that most people, most
of the time, even in periods of violent upheaval, go about their
legitimate business in legitimate ways.

So far there has been no sign of reversal, even stabilisation, in
the trends of crime. For the time being we have to live with it and
try to contain it. For all its imperfections, the criminal law is
designed not merely as a buttress for the privileges of the powerful but as a shield for the elemental human liberties of the poor
and the weak against the assaults of the strong and the treacherous. In that context the rigour of the law must be seen as an
expression of social concern. There is a place for severity of
sentence in response to deliberate and callous crime. But that does
not mean that we must also accept, let alone collude in, the erosion
of criminal justice or deliberate inhumanity in dealing with offenders. To do so is as unlikely as any other approach to bring
about a lasting reduction. It would simply heap other evils on top
of the evil of crime.

Above all, in a world where so few countries still pay more
than lip-service to the ideal of freedom under law, it is vital that
these few do not allow it to become extinguished.

NOTES

Chapter 1

1. National Advisory Commission on Criminal Justice Standards and Goals *Report on Corrections*, Washington, D. C., 1973, p. 352.
2. Bell, Daniel, *The End of Ideology* (Revised Edition), Free Press, New York, 1962, p. 154.
3. Fifth United Nations Congress on the Prevention of Crime and the Treatment of Offenders, Geneva, 1975. Working paper prepared by the Secretariat: *Changes in Forms and Dimensions of Criminality, transnational and national*, pp. 46-7.
4. See also on this, Quetelet, L.A.J., *Physique Sociale* (Revised Edition), Vol. II, Brussels, 1869, pp. 299-308.
5. Alcock, Thomas, *Observations on the Defects of the Poor Laws*, London, 1752, p. 68.
6. *Traité sur la Mendicité avec Projects de Réglement propres à l'empêcher dans les Villes et Villages, Dedié à Messieurs les Officiers de Justice et de Police, par un Citoyen*, Paris, 1774, Preface at p. 1.

Chapter 2

7. Oba, S., *Unverbesserliche Verbrecher und ihre Behandlung*, Berlin, 1908, p. 27.
8. Chadwick, E., "Preventive Police" in *London Review*, Vol. I, 1829, pp. 301-2.

Chapter 3

9. Guerry, A-M., *Statistique morale de l'Angleterre comparée avec la Statistique morale de la France*, Paris, 1864, p. lvii.
10. See Quetelet, L.A.J., 1869, *op. cit.*, p. 317.
11. Lacassagne, A., in *Actes du Premier Congrès International d'Anthropologie Criminelle*, Rome, 1885, p. 167.
12. Durkheim, E., *The Rules of Sociological Method*, 1895, translated by Solovay, S.A., and Mueller, J.M., ed. Carlin, G.E.G., The Free Press, Glencoe, Ill., 1938, reprinted 1958, p. 67.

13. Buchholz, E., Hartmann, R., Lekschas, J., and Stiller, G., *Socialist Criminology*, English translation, D.C. Heath Ltd., Lexington, Mass., 1974, p. 185.

Chapter 4

14. Clark, R., *Crime in America*, Simon & Schuster, New York, 1970, p. 38.

Chapter 5

15. Teaching and Research Office for Criminal Law of the Central Political-Legal Cadres' School (Lectures on the general principles of criminal law of the People's Republic of China), quoted in Cohen, J. A., *The Criminal Process in the People's Republic of China, 1948-1963*, Harvard Univ. Press, Cambridge, Mass., 1968, pp. 328-34, at p. 330.
16. Mill, J.S., *On Liberty* (First published 1859), Pelican Books, 1974. Harmondsworth, Penguin, p. 68.
See also on this, St. John Packe, M., *The Life of John Stuart Mill*, Secker & Warburg, London, 1954, pp. 400-60.

Chapter 7

17. See on this, Myrdal, G., *Asian Drama*, Twentieth Century Fund, New York, Penguin Books, Vol. 2, 1968, pp. 937-58; also Clinard, M.B., and Abbott, D.J., *Crime in Developing Countries*, John Wiley & Sons, New York, 1973, pp. 219-22.

Chapter 8

18. Fitzjames Stephen, Sir James, *A History of the Criminal Law of England*, Macmillan, London, Vol. II, 1883, p. 81.

Chapter 9

19. *Report of the Departmental Committee on Prisons* (known as the Gladstone Committee), 1895, LVI, 1, Minutes of Evidence Nos. 11.482 and 11.483-85, H.M.S.O., London.

20. *Memoirs of the Life of Sir Samuel Romilly* (ed. by his sons), London, Vol. 1, 1840, pp. 169–70.

Chapter 11

21. Martinson, R., "What Works?" in *Journal of Public Interest*, No. 36, June, 1974, p. 22; Lipton, D., Martinson, R., and Wilks, J., *The Effectiveness of Correctional Treatment*, New York: Praeger, 1975; Brody, S.R., *The Effectiveness of Sentencing*, H.M.S.O., London, 1975.

ACKNOWLEDGMENTS

We should like to express our thanks to Mr. George Rainbird, former Deputy Chairman of Thomson Publications, and a publisher in his own right, for his constant encouragement in the writing of this book. We are highly indebted to Mr. Roger Machell of Hamish Hamilton and to Mr. Martin Kessler of Basic Books for their painstaking scrutiny of the text and much editorial guidance.

We owe a great deal to Dr. Joseph Black of the Rockefeller Foundation and to Mr. Christopher Edley and Mr. Sanford Jaffe of the Ford Foundation; they were instrumental in securing a measure of much needed support which was gratefully accepted by the University of Cambridge.

Amongst our colleagues from the Cambridge Institute of Criminology we have been particularly grateful to Dr. Donald West, Reader in Clinical Criminology, for allowing us to draw upon his extensive knowledge of psychiatric and psychological aspects of criminal behaviour. Dr. Roger Hood, now Reader in Criminology at the University of Oxford, has perused all of the chapters and made many helpful suggestions. Dr. Keith Hawkins, now also at Oxford, presented us with a thoughtful memorandum on the Californian system of parole. Mr. Nicholas Miller carried out some tedious but important computations.

Dr. Paul O'Higgins, of the Cambridge Faculty of Law, let us have the benefit of his advice on the topic of international law in relation to crime. Mr. John Gross, Editor of the Times Literary Supplement, gave permission to use, as one of the chapters of the book, a revised version of the article published in the issue of the 14th September, 1975.

Amongst our American friends we would like to single out Professor Herbert Wechsler who has done us the courtesy of reading and discussing with us our comments on sentencing and on parole. Professor Marvin Wolfgang and Professor Peter Low have also helped us on various points.

Lord Allen of Abbeydale (formerly Sir Philip Allen, Permanent Under-Secretary of State at the Home Office) took a great interest in this venture, especially when it was at its planning stage. His successor, Sir Arthur Peterson, looked at some of the sections. Sir Robert Mark and Deputy Commissioner Colin Woods, of Scotland Yard, found time to peruse our chapter on police and supplied additional illustrative material. We are similarly grateful to Mr. Michael Moriarty of the Home Office for his comments upon recent trends of murder in England and for a brief but informative memorandum on the incidence of violence in Northern Ireland. None of these, of course, is responsible in any way for the views expressed in the book.

Many libraries, including that of the United Nations, have been consulted, but we should like to record in particular the help given us by the Library of the Columbia Law School and the Radzinowicz Library of the Institute of Criminology.

It would be very difficult, and indeed very cumbersome, to acknowledge in detail the hundreds of sources on which we have drawn, or to thank individually the very many people around the world who have so generously shared with us their experience and their ideas. But we are indebted to them all.

INDEX